Basics of
Atmospheric Dynamics

R.N. Keshavamurty

CRC Press is an imprint of the
Taylor & Francis Group, an **informa** business

INDIAN METEOROLOGICAL SOCIETY

Mausam Bhavan Complex, Lodi Road,

New Delhi - 110 003.

BSP BS Publications

A unit of **BSP Books Pvt. Ltd.**

4-4-309/316, Giriraj Lane,

Sultan Bazar, Hyderabad - 500 095

First published 2022
by CRC Press
2 Park Square, Milton Park, Abingdon, Oxon, OX14 4RN

and by CRC Press
6000 Broken Sound Parkway NW, Suite 300, Boca Raton, FL 33487-2742

© 2022 BS Publications

CRC Press is an imprint of Informa UK Limited

The right of R.N. Keshavamurthy to be identified as author of this work has been asserted in accordance with sections 77 and 78 of the Copyright, Designs and Patents Act 1988.

Print edition not for sale in South Asia (India, Sri Lanka, Nepal, Bangladesh, Pakistan or Bhutan).

British Library Cataloguing-in-Publication Data
A catalogue record for this book is available from the British Library

Library of Congress Cataloging-in-Publication Data
A catalog record has been requested

ISBN: 978-1-032-14008-7 (hbk)
ISBN: 978-1-003-24611-4 (ebk)

DOI: 10.1201/9781003246114

Foreword

The Indian Meteorological Society (IMS) was established in 1957 as a non-profit scientific society with the basic objectives of advancement of meteorological and allied sciences and dissemination of associated knowledge. Currently, IMS has more than 3500 life members distributed among its 32 local chapters. For the last several years, IMS has been successfully engaged with the society at large in spreading awareness about the weather and climate related issues by organizing public lectures, teachers training, media workshops, students interactive sessions, essay/debate and quiz competitions and other similar events. Sometimes these events are organized in collaboration with other national and international organizations. IMS, being a scientific society, organizes national and international symposia (TROPMET/INTROMET) annually and also seminars at national and regional levels. IMS biannually brings out the society journal VayuMandal (Bulletin of Indian Meteorological Society) which contains review articles by eminent meteorologists, regular scientific papers, report on extreme weather events and report of activities of IMS and its local chapters. Society also publishes reports of symposia, seminars and workshops. In addition to the above activities, as part of the Diamond Jubilee Celebration of IMS, decision was taken to publish Popular Book Series for the benefit of students and the general public, especially on complex meteorological topics. So far three books have been published by IMS on 'Weather Satellites', 'Radio Meteorology' and 'Numerical Weather Prediction'.

It is well known that the atmosphere is a special category of fluid whose density exponentially decreases with height and also it rotates along with the earth about the earth's axis. In the beginning of the last century, it was realized by the scientists that the atmosphere along with its associated physical processes can be represented by a set of well known mathematical equations which eventually can be used to understand the weather systems occurring in the earth-atmosphere-ocean system. These equations are nonlinear in nature and it is very difficult to apply these in forecasting of weather at different time scales. Nevertheless, improvements in observational systems, scientific understanding of physical processes and advancements in computing technology have helped a lot in weather forecasting and climate predictions. Atmospheric dynamics is the basis of weather and climate studies. It is very essential to understand the basics of atmospheric dynamics where the curved surface of the earth and its rotation around its own axis plays very important roles. There are some basic concepts which need clarity at the beginning of any course on numerical weather prediction.

This book on 'Basics of Atmospheric Dynamics' is authored by Dr. R.N. Keshavamurty. IMS is very much thankful to Dr. Keshavamurty for devoting his time to write the book which will eventually benefit the students. He made outstanding contributions in advancing scientific understanding of the Indian Monsoon. He played a key role in developing climate modeling activities in India. He has published several important papers in reputed journals and supervised several students leading to PhD. He authored a book on 'Physics of Monsoons'. After his tenure as Director of IITM, Prof. Keshavamurty devoted several years in lecturing on Atmospheric Science in Bangalore University, S.V.University Tirupati, Allahabad University and in several other institutions. This book which is based on several years of research and teaching experience of Prof. R.N. Keshavamurty, I am sure, will help the students of meteorology to a great extent.

S K Dash
President, IMS

Preface

This book is a result of my lectures at several university departments and SERC advanced training programs. It was felt that if these are compiled and made into a book form, it could be of help to beginners and students. The emphasis is on basic concepts. Thus some details have been pushed to appendices.

I have been fortunate to be a part of great institutions like Indian Institute of tropical Meteorology, Pune (as a research scientist and director), Physical Research Laboratory, Ahmedabad (as a professor), GFDL, Princeton University (as a visitor) and India Meteorological department.

During my research career I benefited greatly by discussions with Prof. S. Manabe, Prof. T. Murakmi, Dr. R. Ananthakrishnan, Dr. Kirk Bryan and Dr. K.M. Lau.

I also recall the kind care of Prof. P.R. Pisharoty, Prof. G.C. Asnani, Prof. S.P. Pandya, Prof. J. Smagorinsky, Prof. D. Lal and Prof. C.S. Ramage.

In International Meteorological Centre, Bombay (now Mumbai), under IIOE, I was exposed to synoptic meteorology. My first introduction to dynamic meteorology was by Sri K.V. Rao in the training division of IMD, Pune. Earlier in Central College, Bengaluru I greatly benefited from a course in mathematical physics (as part of Physics Honours). The teachers in National High School and College instilled a keen interest in Physics.

For illustrating synoptic features I have used the excellent high resolution analysis maps of NCMRWF available on the internet as also INSAT cloud pictures from IMD website. These are gratefully acknowledged. The monsoon depression maps were provided by Dr. Geeta Agnihotri, IMD. This is also gratefully acknowledged. The mean wind maps, velocity potential maps and diabatic heating maps based on the excellent NCEP and ERA reanalysis data as also IITM earth system model simulation maps were kindly provided by Dr. Ayantika Dey Choudhury of IITM. This is gratefully acknowledged. I also wish to thank Sri N. Ramdoss of IMD and Sri Suresh Babu for help with the preparation of the manuscript and also Mr. A. Ravikiran of ASC College for help with the diagrams.

I wish to express my thanks to my daughter Laxmi Rao for help with diagrams and computer graphics. I wish to thank my wife Seetha for taking care of everything in our life.

I wish to thank the Indian Meteorological Society for agreeing to publish this book under their banner.

- R.N.K.

Contents

Chapter 3

Atmospheric General Circulation

Chapter 4

Waves in the Atmosphere

Chapter 5

Large Scale Atmospheric Modeling

Chapter 6

Hydrodynamic Instability

Appendices

1. Basic Governing Equations

1.1 Introduction

Many of the atmospheric motion systems are driven by differential heating. Examples of this are the atmospheric general circulation as also some weather systems. In particular, tropical cyclones are driven by the release of latent heat in the central region. Thus we need equations of fluid dynamics and thermodynamics for the study of such systems.

The basic governing equations of atmospheric motion are:
- (i) Equation of motion
- (ii) Continuity equation
- (iii) Equation of state
- (iv) First law of thermodynamics

1.2 Equation of Motion

This is a statement of Newton's second law of motion for a parcel of air (of unit mass) in the atmosphere. It is written as

$$\frac{d\vec{V}}{dt} = -\alpha \nabla p + \vec{g} - 2\vec{\Omega} \times \vec{V} + \vec{F}$$

where p is pressure, $\alpha = \dfrac{1}{\rho}$, ρ is density, $\vec{V}$ is vector wind, $\vec{g}$ is gravity, $\vec{\Omega}$ is angular velocity of earth, $\vec{F}$ is frictional force.

The acceleration is equal to the vector sum of all forces (per unit mass, as mentioned above) acting on the parcel, such as

- (i) Pressure gradient force[1]
- (ii) Gravity
- (iii) Coriolis force
- (iv) Frictional force

---

[1]Appendix 1

The coriolis force arises from the rotation of the earth. We begin with a non-rotating co-ordinate system fixed at the centre of the earth and then transform to a rotating co-ordinate system[2].

It is more convenient to decompose this vector equation into its three scalar components in a local Cartesian system with the *x*-axis pointing to the east, *y*-axis pointing to the north and vertical axis pointing upwards.

The component scalar equations are

$$\frac{du}{dt} = -\alpha\frac{\partial p}{\partial x} + fv + F_x$$

$$\frac{dv}{dt} = -\alpha\frac{\partial p}{\partial y} - fu + F_y$$

$$\frac{dw}{dt} = -\alpha\frac{\partial p}{\partial z} - g + F_z$$

where *u*, *v*, *w* are the scalar components of the wind $\overrightarrow{V}$ along the *x*, *y* and *z* co-ordinates and $f = 2\Omega\sin\phi$ is known as the coriolis parameter. ϕ is the latitude.

In the vertical equation of motion, for large-scale motion, the vertical acceleration is usually small and there is a very good balance between gravity and the vertical pressure gradient force. This is known as hydrostatic balance.

$$-\alpha\frac{\partial p}{\partial z} = g$$

The total derivative $\frac{du}{dt}$ is a derivative following the fluid parcel.

$$\frac{du}{dt} = \frac{\partial u}{\partial t} + u\frac{\partial u}{\partial x} + v\frac{\partial u}{\partial y} + w\frac{\partial u}{\partial z}$$

where $\frac{\partial u}{\partial t}$ is the local derivative at a location and $u\frac{\partial u}{\partial x}$, $v\frac{\partial u}{\partial y}$ and $w\frac{\partial u}{\partial z}$ are known as advection terms.

---

[2]Appendix 2

1.3 Continuity Equation

This is a statement of the physical principle of conservation of mass.

Consider a fixed volume $dx\,dy\,dz$.

Rate of inflow of fluid at wall $A = \rho u\,dy\,dz$
(Fig. 1.1)

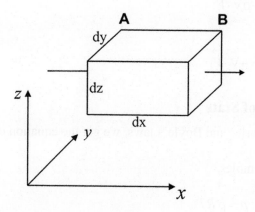

Fig. 1.1

At wall B it is $-\left\{\rho u + \dfrac{\partial}{\partial x}(\rho u)\,dx\right\}dy\,dz$

It is an outflow.

Net inflow $= \rho u\,dy\,dz - \left(\rho u + \dfrac{\partial}{\partial x}(\rho u)\,dx\right)dy\,dz$

$$= -\frac{\partial}{\partial x}(\rho u)\,dx\,dy\,dz$$

Similarly, inflow at all the six walls

$$= -\left\{\frac{\partial}{\partial x}(\rho u) + \frac{\partial}{\partial y}(\rho v) + \frac{\partial}{\partial z}(\rho w)\right\}dx\,dy\,dz$$

This is equal to the local rate of change of mass $\dfrac{\partial}{\partial t}\left(\rho\,dx\,dy\,dz\right)$.

Therefore,

$$\frac{\partial \rho}{\partial t} = -\nabla \cdot \left(\rho \vec{V} \right)$$

$$= -\vec{V} \cdot \nabla \rho - \rho \nabla \cdot \vec{V}$$

i.e., $$\frac{d\rho}{dt} = -\rho \nabla \cdot \vec{V}$$

or $$\frac{1}{\rho} \frac{d\rho}{dt} = -\nabla \cdot \vec{V}$$

1.4 Equation of State

By combining Charles and Boyle's laws, we get the equation of state

$pV = nR_u T$ for n moles.

For unit mass it is $p = \rho RT$

or $p\alpha = RT$ where R (gas constant) is for unit mass.

For the temperature range we encounter in the massive part of the atmosphere, we can treat the atmosphere as a perfect gas (mixture of perfect gases). See Appendix 3.

1.5 First Law of Thermodynamics

If we add a small quantity of heat dQ to a fluid, part of it goes to increase the internal energy (dU) and part of it goes to do work (dW).

Thus, $dQ = dU + dW$

$$dQ = dU + p \, d\alpha \quad \text{for unit mass.}$$

If heat is added at constant volume $d\alpha = 0$.

Then $dQ = dU = C_v dT$

Therefore, in general $dQ = C_v \, dT + p \, d\alpha$

It is to be noted that dQ is not a perfect differential

Now, $p\alpha = RT$

Differentiating

$$p\,d\alpha + \alpha\,dp = R\,dT$$

Therefore, $dQ = C_v\,dT + R\,dT - \alpha\,dp$

Now, if we add heat at constant pressure $dp = 0$

Therefore, $dQ = C_p\,dT = (C_v + R)\,dT$

Therefore in general,

$$dQ = C_p\,dT - \alpha\,dp$$

Writing these as differential equations

$$\dot{Q} = C_v\frac{dT}{dt} + p\frac{d\alpha}{dt}$$

$$= C_p\frac{dT}{dt} - \alpha\frac{dp}{dt}$$

It is convenient to define a new variable called potential temperature.

$$\theta = T\left(\frac{p_0}{p}\right)^{\kappa}$$

where $\kappa = \dfrac{R}{C_p}$ and p_0 is a reference pressure usually taken as 1000 hpa (or 1000 mb).

The potential temperature θ of a parcel at temperature T and pressure p is defined as the temperature attained by it when it is adiabatically brought (compression or expansion) to 1000 hpa. Adiabatic process is one in which heat is neither added nor taken out i.e., $dQ = 0$.

$\dot{Q}$ is rate of heating per unit mass

1.6   Co-ordinate Systems

The most commonly used co-ordinate system is the local co-ordinate system. This is a right-handed Cartesian co-ordinate system on a local tangential plane, which is fixed to the earth at the local point of consideration. *x*-axis points to the east. Distance towards the east is positive and to the west is negative. *y*-axis points to the north. Distance towards north is positive and towards south is negative. *z*-axis points upwards, distance upwards is positive and downwards is negative. (Fig. A2.2)

$\hat{i}, \hat{j}, \hat{k}$ are unit vectors.

The position vector $\vec{r} = x\hat{i} + y\hat{j} + z\hat{k}$

$$\frac{dx}{dt} = u, \quad \frac{dy}{dt} = v, \quad \frac{dz}{dt} = w$$

The velocity (vector)

$$\vec{V} = u\hat{i} + v\hat{j} + w\hat{k}$$

Acceleration,

$$\frac{d\vec{V}}{dt} = \frac{du}{dt}\hat{i} + \frac{dv}{dt}\hat{j} + \frac{dw}{dt}\hat{k}$$

Now, $\quad \dfrac{du}{dt} = \dfrac{\partial u}{\partial t} + u\dfrac{\partial u}{\partial x} + v\dfrac{\partial u}{\partial y} + w\dfrac{\partial u}{\partial z}$

$\dfrac{\partial u}{\partial t}$ is known as the local derivative or local change and $u\dfrac{\partial u}{\partial x}, v\dfrac{\partial u}{\partial y}$ are known as the horizontal advection terms and $w\dfrac{\partial u}{\partial z}$ is known as the vertical advection term.

The physical significance of these terms needs to be understood clearly. It is easier to understand if we use the variable, temperature *T*.

$\dfrac{\partial T}{\partial t}$ is the local change of temperature. If we measure temperature at one location, say the meteorological observatory at Bangalore and note the rate of change of temperature with respect to time at this location, it is known as the local change.

If we have a situation such that isotherms (line connecting places of same temperature) run north-south as shown in the diagram and we have a westerly wind (wind blowing from the west) then air is moving from colder region to a warmer region. This is a case of cold advection. Fig. 1.2(a)

If we have isotherms and wind as in the next diagram, we have a case of warm advection. Fig. 1.2(b).

If we have isotherms running east-west and a southerly wind (wind blowing from the south) as shown in the next diagram, we have cold advection. Fig. 1.2(c).

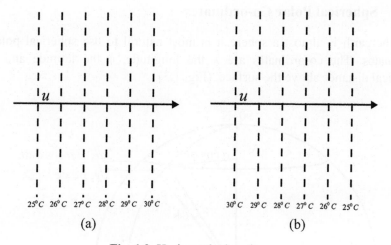

Fig. 1.2 Horizontal advection

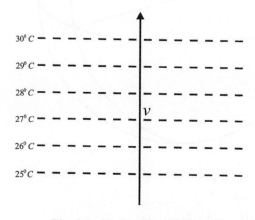

Fig. 1.2c Horizontal advection

Thermal advection is very important in the middle and higher latitudes (extra-tropics). When there are strong northerly winds blowing from Canada towards USA during (northern) winter, it is a case of cold advection. It is sometimes referred to as the Canadian express.

In Asia if there are strong northerly winds (winds blowing from the north) blowing from Siberia in winter, it is a case of cold advection. It can be referred to as the Siberian express. India is protected from these cold winds by the mighty Himalayas.

In the tropics thermal advection is less pronounced. Here moisture advection is prominent. There is large moisture advection by the monsoon south-westerly and southerly winds from the Arabian sea and the Indian ocean.

1.7   Spherical Polar Co-ordinates

As the earth is almost a sphere it is most natural to use spherical polar co-ordinates. The co-ordinates are λ the longitude, ϕ the latitude and z the vertical distance above the surface. (Fig. 1.3).

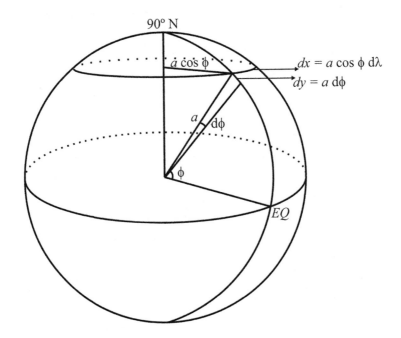

Fig. 1.3 Spherical polar co-ordinates

Now $dx = a\cos\phi\, d\lambda$ and $dy = a\, d\phi$

The velocity is

$$\vec{V} = \hat{i}\,u + \hat{j}\,v + \hat{k}\,w$$

where $u = a\cos\phi\dfrac{d\lambda}{dt}$, $v = a\dfrac{d\phi}{dt}$ and $w = \dfrac{dz}{dt}$

$$r = a + z \approx a$$

where a = radius of the earth.

It is to be noted that the unit vectors $\hat{i}$, $\hat{j}$ and $\hat{k}$ are functions of the position on the earth.

Taking this into consideration, the component equations are

$$\frac{du}{dt} - \frac{uv\tan\phi}{a} + \frac{uw}{a} = -\alpha\frac{\partial p}{\partial x} + 2\Omega v \sin\phi - 2\Omega w\cos\phi + F_x$$

$$\frac{dv}{dt} + \frac{u^2\tan\phi}{a} + \frac{vw}{a} = -\alpha\frac{\partial p}{\partial y} - 2\Omega u\sin\phi + F_y$$

$$\frac{dw}{dt} - \frac{u^2 + v^2}{a} = -\alpha\frac{\partial p}{\partial z} - g + 2\Omega u\cos\phi + F_z$$

For synoptic and large scale motions, the horizontal length scales are of the order of 1000 km, time scale is of the order of a day (10^5 second), vertical height scale is of the order of 10 km and horizontal wind speed is of the order of 10 ms^{-1}. However, the vertical wind speed is much smaller and is of the order of 1 cms^{-1}.

For these scales (or small Rossby number[3] regime), the equations of motion take the simpler form

$$\frac{du}{dt} = -\alpha\frac{\partial p}{\partial x} + fv + F_x$$

$$\frac{dv}{dt} = -\alpha\frac{\partial p}{\partial y} - fu + F_y$$

$$\frac{dw}{dt} = -\alpha\frac{\partial p}{\partial z} - g + F_z$$

---

[3]Appendix 4

For large scale motion, the accelerations are one order of magnitude smaller. So,

$$+\alpha\frac{\partial p}{\partial x} \approx fv \qquad v_g = \frac{\alpha}{f}\frac{\partial p}{\partial x}$$

$$+\alpha\frac{\partial p}{\partial y} \approx -fu \qquad u_g = -\frac{\alpha}{f}\frac{\partial p}{\partial y}$$

This is known as geostrophic balance and $\vec{V}_g = u_g\,\hat{i} + v_g\,\hat{j}$ is the geostrophic wind.

In vector notation

$$-\alpha\nabla_H p - f\,\hat{k} \times \vec{V}_g = 0$$

i.e., $$\vec{V}_g = -\frac{\alpha}{f}\,\nabla_H\,p \times \hat{k}$$

Large scale wind blows parallel to lines of constant pressure or isobars. See Fig. 1.5c.

During the monsoon season the pressure gradient over peninsular India is from south to north. Therefore the zonal wind is westerly (with low pressure to the left in the northern hemisphere).

In addition, we have, of course, the hydrostatic balance i.e.,

$$\alpha\frac{\partial p}{\partial z} = -g$$

1.8   Pressure Co-ordinates

Now, $dp = -g\rho\,dz$

As the large scale atmosphere is in good hydrostatic balance it is possible to use pressure as a vertical co-ordinate (independent variable) instead of height z. Then z, the height of constant pressure surface or $\phi = gz$, the geopotential becomes the dependent variable.

In, x, y, p – co-ordinates; $\dot{x} = u$, $\dot{y} = v$, $\omega = \dfrac{dp}{dt}$.

The total derivatives is

$$\frac{du}{dt} = \frac{\partial u}{\partial t} + u\frac{\partial u}{\partial x} + v\frac{\partial u}{\partial y} + \omega\frac{\partial u}{\partial p}$$

The equations of motion in this co-ordinate system are

$$\frac{du}{dt} = -\frac{\partial \phi}{\partial x} + fv + F_x$$

$$\frac{dv}{dt} = -\frac{\partial \phi}{\partial y} - fu + F_y$$

$$-g\frac{\partial z}{\partial p} = -\frac{\partial \phi}{\partial p} = \frac{1}{\rho} = \alpha \quad \text{or} \quad \alpha = -\frac{\partial \phi}{\partial p}$$

The continuity equation is

$$\frac{\partial u}{\partial x} + \frac{\partial v}{\partial y} + \frac{\partial \omega}{\partial p} = 0$$

Thus it becomes simpler.

1.9 Natural Co-ordinates

We consider a unit vector $\hat{t}$ parallel to the horizontal wind vector (or streamline) and $\hat{n}$ perpendicular to the wind, positive to the left of the flow direction.

$\hat{k}$ is directed vertically upwards.

$$\vec{V} = V\hat{t}$$

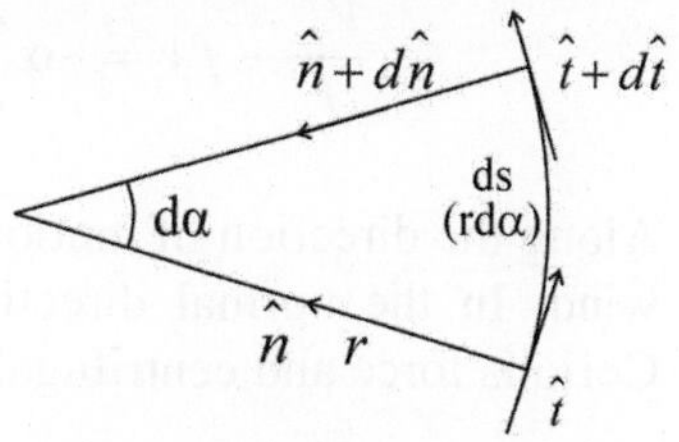

Fig. 1.4 Natural coordinates

$$\frac{d\vec{V}}{dt} = \frac{dV}{dt}\hat{t} + V\frac{d\hat{t}}{dt}$$

From the figure 1.4, $d\hat{t} = d\alpha\hat{n}$

$$\frac{d\vec{V}}{dt} = \dot{V}\hat{t} + V\frac{d\alpha}{dt}\hat{n}$$

$$= \dot{V}\hat{t} + V\frac{1}{r}\frac{rd\alpha}{dt}\hat{n}$$

where r = radius of curvature

$$= \dot{V}\hat{t} + \frac{V^2}{r}\hat{n}$$

$\dot{V} = \dfrac{dV}{dt}$ is the tangential acceleration

$\dfrac{V^2}{r}$ is the centrifugal acceleration

Since the Coriolis force acts always normal to the direction of motion

$$-f\hat{k} \times \vec{V} = -fV\hat{n}$$

The equations of motion in natural co-ordinates are

$$\frac{dV}{dt} = -\alpha\frac{\partial p}{\partial s}$$

$$\frac{V^2}{r} + fV = -\alpha\frac{\partial p}{\partial n}$$

Along the direction of motion, the tangential pressure gradient accelerates the wind. In the normal direction, the pressure gradient force is balanced by Coriolis force and centrifugal force.

1.10 Gradient Wind

Here the flow (wind) is parallel to isobars and is called gradient wind. (Fig. 1.5 a, b)

The tangential acceleration,

$$\frac{dV}{dt} = 0$$

$$\frac{V^2}{r} + fV = -\alpha\frac{\partial p}{\partial n}$$

There is a balance between pressure gradient force, coriolis force and centrifugal force.

In a low pressure area (Fig. 1.5a) the flow is anticlockwise in the northern hemisphere. In regions of high pressure the flow is clockwise. Regions of low pressure or cyclonic circulation in the lower levels are usually associated with weather. An exception is the shallow heat low over northwest India-Pakistan during the monsoon season. Regions of high pressure or anticyclones (Fig. 1.5b) in the lower levels are usually associated with sinking and clear weather.

If $r \to \infty$

The flow is in a straight line

$$fV_g = -\alpha\frac{\partial p}{\partial n}$$

This is geostrophic wind.

There is a balance between pressure gradient force and coriolis force. (Fig. 1.5 c).

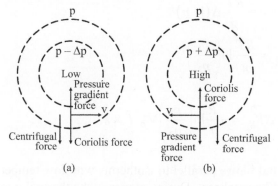

Fig. 1.5 (a & b) Gradient wind

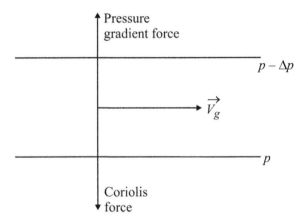

Fig. 1.5 (c) Geostrophic wind

1.11   Thermal Wind-variation of Wind in the Vertical

The geotrophic wind $\vec{V}g = -\dfrac{g}{f}\nabla_p z \times \hat{k}$

In the vertical $\Delta_z \vec{V}g = -\dfrac{g}{f}\nabla_p (\Delta z) \times \hat{k}$

$$dp = -g\rho\,dz$$

$$\alpha\,dp = -g\,dz$$

$$p\alpha = RT$$

$$RT\frac{dp}{p} = -g\,dz = -d\phi$$

$$dz = -\frac{RT}{g}d\ln p$$

$$\Delta z = -\frac{R\overline{T}}{g}\Delta(\ln p)$$

$$= \frac{R\overline{T}}{g}\ln\left(\frac{p_1}{p_2}\right)$$

Thermal wind $\Delta\vec{V}g = -\dfrac{R}{f}\ln\left(\dfrac{p_1}{p_2}\right)\nabla_p \overline{T} \times \hat{k}$

The thermal wind blows parallel to isotherms with low temperature to the left (in the northern hemisphere). During monsoon the thermal wind is easterly.

1.12 Vorticity and Divergence

There are two quantities derived from the wind $\vec{V}$, which are very useful.

Vorticity $\quad \nabla \times \vec{V} = \xi \hat{i} + \eta \hat{j} + \zeta \hat{k}$

The vertical component of vorticity is

$$\zeta = \hat{k} \cdot \nabla \times \vec{V} = \frac{\partial v}{\partial x} - \frac{\partial u}{\partial y}$$

The circulation around $ABCDA$ (Fig. 1.6)

$$= \oint \vec{V} \cdot dl$$

$$dC = \left(u - \frac{\partial u}{\partial y}\frac{dy}{2} \right)dx + \left(v + \frac{\partial v}{\partial x}\frac{dx}{2} \right)dy$$

$$- \left(u + \frac{\partial u}{\partial y}\frac{dy}{2} \right)dx - \left(v - \frac{\partial v}{\partial x}\frac{dx}{2} \right)dy$$

$$= u\,dx - \frac{\partial u}{\partial y}\frac{dy}{2}dx + v\,dy + \frac{\partial v}{\partial x}\frac{dx}{2}dy$$

$$- u\,dx - \frac{\partial u}{\partial y}\frac{dy}{2}dx - v\,dy + \frac{\partial v}{\partial x}\frac{dx}{2}dy$$

$$= \left(\frac{\partial v}{\partial x} - \frac{\partial u}{\partial y} \right)dx\,dy$$

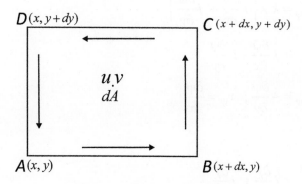

Fig. 1.6 Circulation and vorticity

The circulation per unit area

$$\zeta = \frac{dC}{dA} = \frac{\partial v}{\partial x} - \frac{\partial u}{\partial y}$$

Vorticity shows rotation,

Divergence

$$D = \nabla \cdot \vec{V} = \left(\hat{i} \frac{\partial}{\partial x} + \hat{j} \frac{\partial}{\partial y} + \hat{k} \frac{\partial}{\partial y} \right) \cdot \left(\hat{i} u + \hat{j} v + \hat{k} w \right)$$

$$= \left(\frac{\partial u}{\partial x} + \frac{\partial v}{\partial y} + \frac{\partial w}{\partial z} \right)$$

Horizontal divergence

$$D_H = \frac{\partial u}{\partial x} + \frac{\partial v}{\partial y}$$

Now the horizontal wind can be expressed as

$$\vec{V} = \hat{k} \times \nabla \psi + \nabla \chi$$

$$= \vec{V}_\psi + \vec{V}_\chi$$

where, ψ - the stream function

χ - velocity potential

$$\vec{V}_\psi = \hat{k} \times \left(\hat{i} \frac{\partial \psi}{\partial x} + \hat{j} \frac{\partial \psi}{\partial y} \right)$$

$$= \hat{j} \frac{\partial \psi}{\partial x} - \hat{i} \frac{\partial \psi}{\partial y}$$

$$u_\psi = -\frac{\partial \psi}{\partial y} ; \quad v_\psi = \frac{\partial \psi}{\partial x}$$

Similarly,

$$\vec{V}_\chi = \hat{i} \frac{\partial \chi}{\partial x} + \hat{j} \frac{\partial \chi}{\partial y}$$

$$u_\chi = \frac{\partial \chi}{\partial x} ; \quad v_\chi = \frac{\partial \chi}{\partial y}$$

Therefore,

$$u = u_\psi + u_\chi$$

$$= -\frac{\partial \psi}{\partial y} + \frac{\partial \chi}{\partial x}$$

$$v = \frac{\partial \psi}{\partial x} + \frac{\partial \chi}{\partial y}$$

The vertical component of vorticity

$$\zeta = \frac{\partial v}{\partial x} - \frac{\partial u}{\partial y} = \frac{\partial}{\partial x}\left(\frac{\partial \psi}{\partial x} + \frac{\partial \chi}{\partial y}\right) - \frac{\partial}{\partial y}\left(-\frac{\partial \psi}{\partial y} + \frac{\partial \chi}{\partial y}\right)$$

$$= \frac{\partial^2 \psi}{\partial x^2} + \frac{\partial^2 \psi}{\partial y^2} = \nabla^2 \psi$$

Only ψ component (the rotational part of wind) contributes to vorticity.

The horizontal divergence

$$D_H = \frac{\partial u}{\partial x} + \frac{\partial v}{\partial y} = \frac{\partial}{\partial x}\left(-\frac{\partial \psi}{\partial y} + \frac{\partial \chi}{\partial x}\right) + \frac{\partial}{\partial y}\left(\frac{\partial \psi}{\partial x} + \frac{\partial \chi}{\partial y}\right)$$

$$= \frac{\partial^2 \chi}{\partial x^2} + \frac{\partial^2 \chi}{\partial y^2} = \nabla^2 \chi$$

Only χ component of wind contributes to divergence.

1.13 Vorticity Equation

From the two component equations of motion along the x (longitude) and y (latitude) directions, it is convenient to obtain a derived equation known as the vorticity equation.

$$\frac{\partial u}{\partial t} + u\frac{\partial u}{\partial x} + v\frac{\partial u}{\partial y} + w\frac{\partial u}{\partial z} = -\alpha\frac{\partial p}{\partial x} + fv + F_x \qquad \text{... (1.1)}$$

$$\frac{\partial v}{\partial t} + u\frac{\partial v}{\partial x} + v\frac{\partial v}{\partial y} + w\frac{\partial v}{\partial z} = -\alpha\frac{\partial p}{\partial y} - fu + F_y \qquad \text{... (1.2)}$$

Taking $\dfrac{\partial}{\partial y}$ of equation (1.1) and subtracting from $\dfrac{\partial}{\partial x}$ of equation (1.2)

$$-\left[\frac{\partial}{\partial t}\left(\frac{\partial u}{\partial y}\right)+\frac{\partial u}{\partial y}\frac{\partial u}{\partial x}+u\frac{\partial}{\partial y}\left(\frac{\partial u}{\partial x}\right)+\frac{\partial v}{\partial y}\frac{\partial u}{\partial y}+v\frac{\partial}{\partial y}\left(\frac{\partial u}{\partial y}\right)+\right.$$

$$\frac{\partial w}{\partial y}\frac{\partial u}{\partial z}+w\frac{\partial}{\partial y}\left(\frac{\partial u}{\partial z}\right)=-\frac{\partial \alpha}{\partial y}\frac{\partial p}{\partial x}-\alpha\frac{\partial}{\partial y}\left(\frac{\partial p}{\partial x}\right)+f\frac{\partial v}{\partial y}+$$

$$\left.\beta v+\frac{\partial F_x}{\partial y}\right] \qquad \dots (1.3)$$

$$+\left[\frac{\partial}{\partial x}\left(\frac{\partial v}{\partial t}\right)+\frac{\partial u}{\partial x}\frac{\partial v}{\partial x}+u\frac{\partial}{\partial x}\left(\frac{\partial v}{\partial x}\right)+\frac{\partial v}{\partial x}\frac{\partial v}{\partial y}+v\frac{\partial}{\partial x}\left(\frac{\partial v}{\partial y}\right)+\right.$$

$$\frac{\partial w}{\partial x}\frac{\partial v}{\partial z}+w\frac{\partial}{\partial x}\left(\frac{\partial v}{\partial z}\right)=-\frac{\partial \alpha}{\partial x}\frac{\partial p}{\partial y}-\alpha\frac{\partial}{\partial x}\left(\frac{\partial p}{\partial y}\right)-f\frac{\partial u}{\partial x}+$$

$$\left.\frac{\partial F_y}{\partial x}\right] \qquad \dots (1.4)$$

Regrouping and adding terms of similar type

$$\frac{\partial}{\partial t}\left(\frac{\partial v}{\partial x}-\frac{\partial u}{\partial y}\right)+u\frac{\partial}{\partial x}\left(\frac{\partial v}{\partial x}-\frac{\partial u}{\partial y}\right)+v\frac{\partial}{\partial y}\left(\frac{\partial v}{\partial x}-\frac{\partial u}{\partial y}\right)+$$

$$w\frac{\partial}{\partial z}\left(\frac{\partial v}{\partial x}-\frac{\partial u}{\partial y}\right)+\left(\frac{\partial u}{\partial x}+\frac{\partial v}{\partial y}\right)\left(\frac{\partial v}{\partial x}-\frac{\partial u}{\partial y}\right)+$$

$$\left(\frac{\partial w}{\partial x}\frac{\partial v}{\partial z}-\frac{\partial w}{\partial y}\frac{\partial u}{\partial z}\right)=-\left(\frac{\partial \alpha}{\partial x}\frac{\partial p}{\partial y}-\frac{\partial \alpha}{\partial y}\frac{\partial p}{\partial x}\right)-f\left(\frac{\partial u}{\partial x}+\frac{\partial v}{\partial y}\right)-$$

$$\beta v+\frac{\partial F_y}{\partial x}-\frac{\partial F_x}{\partial y}$$

i.e.,

$$\frac{\partial \zeta}{\partial t}+u\frac{\partial \zeta}{\partial x}+v\frac{\partial \zeta}{\partial z}+w\frac{\partial \zeta}{\partial z}+\beta v$$

$$=-(\zeta+f)D_H+\left(\frac{\partial w}{\partial y}\frac{\partial u}{\partial z}-\frac{\partial w}{\partial x}\frac{\partial v}{\partial z}\right)+\left(\frac{\partial p}{\partial x}\frac{\partial \alpha}{\partial y}-\frac{\partial p}{\partial y}\frac{\partial \alpha}{\partial x}\right)+$$

$$\left(\frac{\partial F_x}{\partial x}-\frac{\partial F_y}{\partial y}\right)$$

i.e.,
$$\frac{d}{dt}(\zeta+f)=-(\zeta+f)D_H+\left(\frac{\partial w}{\partial y}\frac{\partial u}{\partial z}-\frac{\partial w}{\partial x}\frac{\partial v}{\partial z}\right)+$$
$$\left(\frac{\partial p}{\partial x}\frac{\partial \alpha}{\partial y}-\frac{\partial p}{\partial y}\frac{\partial \alpha}{\partial x}\right)+\left(\frac{\partial F_x}{\partial x}-\frac{\partial F_y}{\partial y}\right)$$

The absolute vorticity of a fluid parcel can change by (i) divergence term, (ii) tilting term, (iii) solenoidal term and, (iv) friction term.

The largest term is the divergence term. In regions of horizontal convergence (D_H negative), the vorticity increases.

The second term is called the vortex tube term. This term converts horizontal component of vorticity into the vertical component.

The third term is called the solenoidal term. If iso-lines of α (or ρ) and iso-lines of pressure or isobars intersect at an angle, we have pressure-specific volume solenoids.

As an approximation

$$\frac{d\zeta_a}{dt}\approx-\zeta_a D_H \quad \text{or} \quad \frac{dl_n}{dt}\zeta_a \approx -D_H = -\nabla_H\cdot\vec{V}$$

If the divergence is zero, as in a nondivergent barotropic atmosphere, then

$$\frac{d\zeta_a}{dt}=0$$

Thus, in this case the absolute vorticity is conserved.

1.14 Humidity

The atmosphere contains a variable amount of moisture. Water occurs in all three phases – ice/snow, water and vapour.

Water vapour is a very important constituent of the atmosphere. Obviously, it is an essential ingredient for cloud and rain formation. It also absorbs terrestrial infrared radiation and keeps the earth warm.

When water vapour condenses to water the released latent heat is an important secondary source of heating for tropical and monsoon disturbances as well as for the maintenance of monsoon circulation and the Hardley circulation.

There is always evaporation from the oceans and other water bodies. The amount of evaporation depends upon wind speed and the vertical gradient of water vapour above the water surface.

The partial pressure of water vapour is referred to as vapour pressure, e. Under equilibrium conditions, when the maximum possible water vapour has evaporated, the atmosphere is said to be saturated. Then the vapour pressure is referred to as saturation vapour pressure, e_s. The mixture of air and water vapour, when it is saturated is referred to as saturated air. When vapour pressure is less than saturation vapour pressure, air is said to be unsaturated. When saturated air is cooled – as it happens when air rises:

(i)   when air flows over a mountain barrier or
(ii) in regions of low level convergence as in low pressure areas, depressions or storms, the saturation vapour pressure at the new lower temperature is less. The atmosphere can hold less amount of water. The excess water vapour usually condenses.

The latent heat of vaporization is the amount of heat required to change unit mass (one gram or kilogram) of water into water vapour at the same temperature. On the other hand when one gram or kilogram of water vapour condenses to water, latent heat is released.

The equation of state for water vapour is

$$e = \rho_v R_v T$$

where R_v is gas constant of water vapour.

Clausius-Clapeyron equation

$$\frac{1}{e_s} \frac{de_s}{dT} = \frac{L_{lv}}{R_v T^2}$$

This tells us about the variation of saturation vapour pressure with temperature. Saturation vapour pressure increases with temperature. At higher temperatures as in the tropics, saturation vapour pressure is higher. The atmosphere can hold more moisture.

As said earlier moist air is a mixture of dry air and water vapour. There are different ways of expressing the moisture content in the atmosphere.

Mixing ratio is the mass of water vapour per unit mass of dry air in the mixture.

$$m = \frac{M_v}{M_d} = \frac{\dfrac{M_v}{V}}{\dfrac{M_d}{V}} = \frac{\rho_v}{\rho_d}$$

Specific humidity q is the mass of water vapour per unit mass of (moist) air.

$$q = \frac{M_v}{M} = \frac{M_v}{M_d + M_v} = \frac{\rho_v}{\rho_d + \rho_v}$$

m and q are dimensionless. They are expressed in gram per kilogram.

Relative humidity is the ratio of the observed mixing ratio and the saturation mixing ratio.

$$r = \frac{m}{m_s} \approx \frac{e}{e_s}$$

Relative humidity is generally expressed as a percentage.

1.15 Hydrostatic Stability

When a small parcel of air is displaced vertically (without any mixing), if it comes back to its original position, then the atmosphere is said to be in hydrostatic equilibrium. If the parcel moves away, the atmosphere is unstable. If the parcel rests in any position it is moved to, it is in neutral equilibrium.

The surrounding air is in hydrostatic equilibrium

i.e., $dp = -g\,\rho\,dz$ ... (1.5)

The parcel (of density ρ') may not be in hydrostatic equilibrium.

$$\frac{dw}{dt} = -g - \frac{1}{\rho'}\left(\frac{\partial p}{\partial z}\right)$$ ... (1.6)

ρ' is the density of displaced parcel, $p' = p$ pressure adjusts quickly.

Eliminate $\dfrac{\partial p}{\partial z}$

$$\frac{dw}{dt} = -g + \frac{1}{\rho'} g\rho = g\left(\frac{\rho - \rho'}{\rho'}\right) \qquad \qquad \ldots (1.7)$$

$$= \frac{g(\gamma - \gamma')z}{T}$$

where $\gamma = -\dfrac{\partial T}{\partial z}$ is the lapse rate

If $\gamma > \gamma'$ Unstable

If $\gamma = \gamma'$ Neutral

If $\gamma < \gamma'$ Stable

The displacements are adiabatic.

In the atmosphere which contains moisture,

$\gamma < \gamma_s$ absolutely stable

$\gamma_s < \gamma < \gamma_d$ conditionally unstable, stable till the parcel is unsaturated, unstable when it becomes saturated.

$\gamma > \gamma_d$ absolutely unstable[4]

γ_d is dry adiabatic lapse rate (DALR) and γ_s is saturation adiabatic lapse rate (SALR).

Now, $\qquad \theta = T\left(\dfrac{p_0}{p}\right)^{R/C_p}$

$$\frac{\partial \ln \theta}{\partial z} = \frac{\partial \ln T}{\partial z} - \frac{R}{C_p} \frac{\partial \ln p}{\partial z}$$

---

[4]Appendix 5

$$\frac{1}{\theta}\frac{\partial \theta}{\partial z} = \frac{1}{T}\frac{\partial T}{\partial z} - \frac{R}{C_p}\frac{1}{p}\frac{\partial p}{\partial z}$$

$$= \frac{1}{T}\frac{\partial T}{\partial z} + \frac{R}{C_p}\frac{g\rho}{\rho RT}$$

$$\frac{1}{\theta}\frac{\partial \theta}{\partial z} = \frac{1}{T}(-\gamma) + \frac{g}{C_p T}$$

or $\qquad \dfrac{T}{\theta}\dfrac{\partial \theta}{\partial z} = \gamma_d - \gamma$

For dry or unsaturated air

$$\frac{\partial \theta}{\partial z} < 0 \quad \text{Unstable}$$

$$\frac{\partial \theta}{\partial z} = 0 \quad \text{Neutral}$$

$$\frac{\partial \theta}{\partial z} > 0 \quad \text{Stable}$$

2. Energetics of the Atmosphere

2.1 Introduction

The earth's atmosphere can be looked upon as a heat engine in which part of the radiant energy received from the sun is converted into the mechanical energy of the winds. The details of this conversion have been the subject of extensive study. In a heat engine we need (i) a source of heat, (ii) a sink of heat and (iii) a working substance, Several traditional and satellite studies have shown that the overall radiation balance (energy received-energy lost) of the earth-atmosphere system is positive in the tropics and negative in the higher latitudes (Fig. 3.2). Thus, we have a source of heat in the tropics, a sink in the higher latitudes and the working substance is the atmospheric air. This heat engine is under the influence of earth's rotation and gravity.

The details of the energy conversion in this heat engine have been among the main problems of general circulation studies. If the global winds are averaged in time and around latitude circles, the most prominent feature noticed is the westerly jet stream in the upper troposphere of middle latitudes (Fig. 3.3, Fig. 3.4). The problem as to how this westerly jet results from the north-south temperature contrast has been the subject of observational and theoretical studies as well as laboratory and numerical experiments.

The importance of large-scale eddies in the transport of momentum and sensible heat was pointed out by Jeffreys (1926). Observational studies of Starr and White (1951) confirmed the important role of large scale eddies in the poleward transport of zonal momentum for the maintenance of the zonal winds in the middle latitude. The theoretical studies of baroclinic and barotropic instabilities (Charney, 1947; Kuo, 1949 and others) added significantly to our understanding of the role of large-scale eddies. The basic framework for the studies of energy conversions in the atmosphere was formulated by Lorenz (1955) and Phillips (1954, 1956). The laboratory experiments of Fultz (1951) and Hide (1953), incorporating differential heating and rotation, produced flow patterns similar to those occurring in the atmosphere.

Several sophisticated and advanced general circulation model studies starting with the pioneering works of Smagorinsky et.al. (1965) and Manabe et.al. (1967) have succeeded in simulating the important features of the observed general circulation.

Recently energetics calculations have been revisited using the excellent reanalysis data (Y.H. Kim, M.K. Kim, 2013; C.A.F. Marques, A. Rocha, J. Cort e-Real, J.M. Caslanheira, 2009).

The consensus of the various observational and theoretical studies and numerical simulations about the middle latitude zonal westerlies is that large-scale eddies grow by conversion from zonal available potential energy (which is converted into eddy available potential energy) and transport zonal momentum towards the zonal maximum wind and thus maintain the westerly jet. In the tropics, however, the Hadley cell plays an important role.

2.2 Energy Equations

The equations governing the gross behaviour of the atmosphere are known from classical physics. The different forms of energy relevant for atmospheric energetics are:

(a) The kinetic energy per unit mass of air is

$$k = \frac{1}{2}\left(u^2 + v^2 + w^2\right)$$

Integrating this over a mass M we get,

$$K = \int_M \frac{u^2 + v^2 + w^2}{2} \rho \, d\tau$$

This can be split up into

(i) the horizontal kinetic energy
$$K_H = \int \frac{u^2 + v^2}{2} \rho \, d\tau$$

(ii) the vertical energy
$$K_w = \int \rho \frac{w^2}{2} \, d\tau$$

(b) the internal energy
$$I = \int c_v T \rho \, d\tau$$

and

(c)   the potential energy

$$P = \int g z \rho \, d\tau$$

The equations, for the (time) rate of change of these forms of energy can be written as (Wiin-Nielsen, 1968):

$$\frac{\partial K_H}{\partial t} = -\int \vec{V_2} \cdot \nabla_2 \, p \, d\tau + \int \vec{V_2} \cdot \vec{F} \rho \, d\tau -$$

$$\int \nabla_3 \cdot \left(\rho \frac{u^2 + v^2}{2} \right) \vec{V_3} \, d\tau \qquad \dots (2.1)$$

$$\frac{\partial I}{\partial t} = \int \rho \dot{Q} d\tau + \int V_2 \cdot \nabla_2 \, p \, d\tau + \int w \frac{\partial p}{\partial z} d\tau -$$

$$\int \nabla_3 \cdot \left(\rho C_V T \vec{V_3} \right) d\tau - \int \nabla_3 \cdot \left(p \vec{V_3} \right) d\tau \qquad \dots (2.2)$$

$$\frac{\partial K_w}{\partial t} = -\int w \frac{\partial p}{\partial z} d\tau - \int g \rho w d\tau -$$

$$\int \nabla_3 \cdot \left(\rho \frac{w^2}{2} \vec{V_3} \right) d\tau \qquad \dots (2.3)$$

and $\quad \dfrac{\partial P}{\partial t} = \int g \rho w \, d\tau - \int \nabla_3 \cdot \left(\rho g z \vec{V_3} \right) d\tau \qquad \dots (2.4)$

We notice that pairs of these equation (2.1, 2.2, 2.3 and 2.4) contain terms equal in magnitude but opposite in sign which can be interpreted as the rate of conversion between the different forms of energy and can be written as:

$$\left. \begin{array}{l} C\left(I, K_H \right) = -\int \vec{V_2} \cdot \nabla_2 \, p \, d\tau \\[2mm] C\left(I, K_w \right) = -\int w \dfrac{\partial p}{\partial z} d\tau \\[2mm] C\left(P, K_w \right) = -\int g \rho w \, d\tau \end{array} \right\} \qquad \dots (2.5)$$

Besides these conversions, all the energy equations contain terms denoting the fluxes of these forms of energy at the boundary.

Subscript 2 indicates two-dimensional
Subscript 3 indicates three-dimensional
$\dot{Q}$ is rate of heating per unit mass

2.3   Physical Explanation of the Energy Conversion Processes

When air in the earth's atmosphere is heated, part of this energy goes to increase its temperature and part of it goes to expand the gas. While expanding, the gas does work. The work done in the horizontal direction produces horizontal kinetic energy and the work done in the vertical direction produces vertical kinetic energy. In the horizontal, the pressure gradient force is roughly balanced by the Coriolis force and the production of horizontal kinetic energy is accomplished by the residual ageostrophic wind i.e., air moving across isobars from high to low pressure. In the vertical, the force of gravity very nearly balances the vertical pressure gradient force and the work of expansion by the small residual non-hydrostatic force produces vertical kinetic energy. The source for both horizontal and vertical kinetic energy is the internal energy. The vertical kinetic energy communicates with potential energy also. The horizontal kinetic energy does not directly communicate with potential energy. Wiin-Nielsen (1968) showed that the assumption of hydrostatic balance makes a difference. The vertical kinetic energy does not grow and that produced by the work done by expansion in the vertical gets converted into potential energy (as the air is raised to higher levels in the earth's gravitational field). Now the horizontal and vertical motions of the atmospheric fluid are related by the equation of continuity, so that potential energy indirectly contributes to horizontal kinetic energy.

2.4   Total Potential Energy (TPE)

If we make the very reasonable hydrostatic assumption, the potential energy

$$P = \int RT\ dm$$

And the internal energy is

$$I = \int C_v T\ dm$$

The potential energy of a column of the atmosphere bears a constant ratio to the internal energy, so that the two can be combined to give the total potential energy (TPE).

$$P + I = \int (C_v + R) T\ dm = \int C_p T\ dm$$

In pressure co-ordinates, the system of equations (2.1) to (2.4) reduce to[1],

---

[1]Appendix 6

$$\frac{\partial K_H}{\partial t} = -\int \vec{V} \cdot \nabla \phi \, dm + \int \vec{V} \cdot \vec{F} \, dm -$$

$$\int \nabla_3 \cdot \left(\vec{V_3} \frac{u^2 + v^2}{2} \right) dm \qquad \ldots (2.6)$$

and

$$\frac{\partial (P + I)}{\partial t} = \int \dot{Q} \, dm + \int \alpha \omega \, dm -$$

$$\int \nabla_3 \cdot \left(\vec{V_3} \, C_p \, T \right) dm \qquad \ldots (2.7)$$

Now,

$$-\int \vec{V} \cdot \nabla \phi \, dm = -\int \nabla \cdot \left(\vec{V} \phi \right) dm + \int \phi \nabla \cdot \vec{V} \, dm$$

$$= -\int \nabla \cdot \left(\vec{V} \phi \right) dm - \int \frac{\partial}{\partial p} (\omega \phi) \, dm - \int \omega \alpha \, dm \qquad \ldots (2.8)$$

If we are considering a closed region

$$-\int \vec{V} \cdot \nabla \phi \, dm = -\int \omega \alpha \, dm$$

Therefore,

$$C \left(TPE, \, K_H \right) = -\int \omega \alpha \, dm \qquad \ldots (2.9)$$

The average value of this per unit area

$$\overline{\overline{C}} \left(TPE, \, K_H \right) = -\int \overline{\overline{\omega \alpha}} \, \frac{dp}{g}$$

$$= -\int \overline{\overline{\omega'' \alpha''}} \, \frac{dp}{g}$$

$$= -R \int \frac{\overline{\overline{\omega'' T''}}}{p} \, \frac{dp}{g} \qquad \ldots (2.10)$$

since $\overline{\overline{\omega}}$ is zero. The double bar denotes an average over an entire isobaric surface and the double prime the deviation from this average. Therefore there is conversion from total potential into kinetic energy when there is horizontal temperature gradient and warm air rises and cold air sinks.

2.5 Available Potential Energy (APE)

According to the laws of thermodynamics, whereas any amount of mechanical energy can be converted into heat by friction, there is a limit to the conversion of heat into mechanical energy. For the latter conversion we need a source of heat and sink of heat (differential heating). The maximum efficiency (η) for conversion of heat into mechanical energy by an ideal reversible heat engine is given by

$$\eta = \frac{W}{Q_1} = \frac{Q_1 - Q_2}{Q_1} = \frac{T_1 - T_2}{T_1}$$

where Q_1 and Q_2 are the heat taken from the source and given to the sink respectively at temperatures T_1 and T_2 and W in the mechanical work done. This is equal to unity only when the sink is at the temperature of absolute zero.

Lorenz (1955) showed that all the total potential energy is not available for conversion into kinetic energy. He considered a reference or ground state of the atmosphere in which all the mass is isentropically stratified. The TPE in this ground state is not available for conversion into kinetic energy as there is no horizontal temperature gradient. In any other state of the atmosphere only that part of TPE which is in excess of that of the ground state is available for conversion and is called the available potential energy. The average value of this (after Lorenz) is

$$\overline{\overline{A}} = \int \frac{\overline{\alpha''^2}}{2\sigma} \frac{dp}{g} \qquad \text{where } \sigma = -\frac{\alpha}{\theta} \frac{\partial \theta}{\partial p} \qquad \qquad \dots (2.11)$$

This is proportional to the variance of specific volume or temperature or potential temperature on a constant pressure surface.

Ven Mieghem (1956) obtained an expression for APE by a different (variational) method. Pfeffer et al (1965) showed that available potential energy depends upon the rate of rotation also. Dutton and Johnson (1967) gave a rigorous derivation of APE. However, as shown by Wiin-Nielsen (1968) the approximate formulation of APE by Lorenz is exact in the quasi-geostrophic model.

2.6 Zonal and Eddy Forms of Energy

The kinetic energy and the available potential energy are customarily split up into zonal and eddy parts as follows:

$$\left[u^2\right] = [u]^2 + \left[u^{*2}\right]$$

where $\qquad [u] = \dfrac{1}{2\pi} \displaystyle\int_0^{2\pi} u \, d\lambda \quad$ and $\quad u^* = u - [u]$

The zonal kinetic energy

$$K_z = \int \frac{[u]^2}{2} \, dm$$

The eddy kinetic energy

$$K_E = \int \frac{\left[u^{*2} + v^{*2}\right]}{2} \, dm$$

The zonal available potential energy

$$A_z = \int \frac{[\alpha]''^2}{2\sigma} \, dm$$

The eddy available potential energy

$$A_E = \int \frac{\left[\alpha^{*2}\right]}{2\sigma} \, dm$$

The equations for the rate of change of these forms of energy are:

$$\frac{\partial K_z}{\partial t} = \frac{\partial}{\partial t} \int \frac{1}{2}[u]^2 \, dm$$

$$= \int [u^* v^*] \frac{\partial}{\partial y}[u] - \int [\omega]'' [\alpha]'' \, dm \qquad \dots (2.12)$$

$$\frac{\partial A_z}{\partial t} = \frac{\partial}{\partial t} \int \frac{1}{2\sigma} [\alpha]''^2 \, dm$$

$$= R^2 \int \frac{1}{\sigma p^2} [v^* T^*] \frac{\partial [T]}{\partial y} \, dm + \int [\omega]'' [\alpha]'' \, dm \qquad \dots (2.13)$$

$$\frac{\partial K_E}{\partial t} = \frac{\partial}{\partial t} \int \frac{1}{2} \left[u^{*2} + v^{*2} \right] dm$$

$$= -\int [u^* v^*] \frac{\partial [u]}{\partial y} \, dm - \int [\omega^* \alpha^*] \, dm \qquad \dots (2.14)$$

$$\frac{\partial A_E}{\partial t} = \frac{\partial}{\partial t} \int \frac{1}{2\sigma} \left[\alpha^{*2} \right] dm$$

$$= -R^2 \int \frac{1}{\sigma p^2} [v^* T^*] \frac{\partial [T]}{\partial y} \, dm + \int [\omega^* \alpha^*] \, dm \qquad \dots (2.15)$$

The inter conversions of these energies are

$$C(A_Z, K_Z) = -\int [\omega]'' [\alpha]'' \, dm \qquad \dots (2.16)$$

$$C(K_E, K_Z) = +\int [u^* v^*] \frac{\partial [u]}{\partial y} \, dm \qquad \dots (2.17)$$

$$C(A_Z, A_E) = -R^2 \int \frac{1}{\sigma p^2} [v^* T^*] \frac{\partial [T]}{\partial y} \, dm \qquad \dots (2.18)$$

$$C(A_E, K_E) = -\int [\omega^* \alpha^*] \, dm \qquad \dots (2.19)$$

The zonal kinetic energy can increase or be maintained against friction by the following processes:

(i) Conversion from ZAPE by *y-p* over turnings in the mean meridional circulation. If the meridional cell is direct i.e., if there is rising of warm air and sinking of cold air in the *y-p* plane there is production of zonal kinetic energy.

(ii) Conversion from eddy kinetic energy. This depends upon the meridional transport of momentum by the eddies. If the eddies transport zonal momentum towards regions of stronger zonal wind (counter-gradient flux) there is production of zonal kinetic energy.

The meridional transport of momentum is related to the tilt of the troughs (eddies) in the horizontal plane. A NE-SW tilted/NW-SE tilted trough transports westerly momentum northward/southward.

Now let us consider the source for the kinetic energy of the eddies. If the troughs (and ridges) are tilted NW-SE to the south of the zonal wind maximum and NE-SW to the north of the maximum, the eddies transport momentum away from the maximum (down gradient flux) and gain kinetic energy at the expense of zonal motion.

The other source for the growth of disturbances is conversion from eddy available potential energy. This is accomplished by over-turnings in the *x-p* plane i.e., rising of warm air and sinking of cold air in the *x-p* plane.

The source for eddy available potential energy is the zonal available potential energy. If the eddies transport sensible heat meridionally down the gradient of temperature, there is conversion from ZAPE to EAPE. If the transport is counter-gradient, there is conversion from EAPE to ZAPE. The meridional transport of sensible heat by eddies is related to the vertical tilt of the disturbance (in the *x-p* plane). A trough in westerlies tilting westwards/ eastwards with height transports sensible heat northwards/ southwards.

Another important source of energy for disturbances in the tropics is the release of latent heat of condensation. The tropical atmosphere is conditionally unstable. Potential temperature increases with height; but equivalent potential temperature decreases with height in the lower troposphere (Rao, 1958). This 'gravitational instability' leads to growth of cumulus scale motion only. Charney and Eliassen (1964) postulated that the large scale disturbances grow with the co- operation of the cumulus scale. The large scale disturbance supplies moisture to cumulus convection by low level frictional convergence and cumulus convection provides released latent heat and produces eddy available potential energy which acts as the source for large-scale eddy kinetic energy. Charney and Eliassen (1964) showed that this mechanism does lead to amplification of the large-scale disturbance. They showed that this effect acts as a phenomenon of instability and called it conditional instability of the second kind (CISK). Here the large-scale motion is forced by the cumulus heating.

2.7   Energy Conversions in the Atmosphere

In the extra tropics the general flow of energy is as shown in Fig. 2.1 (energy conversions in the extra-tropical atmosphere).

The eddies produce EAPE from ZAPE by down gradient flux of sensible heat and convert EAPE into EKE by *x-p* overturnings by the process of baroclinic instability. These eddies transport momentum against the gradient of zonal wind and convert EKE into ZKE. This is the main source of energy for the maintenance of the westerly jet. Starr (1968) called this the phenomenon of 'negative eddy viscosity'. This is not known in classical turbulence theories. Such flow of energy from smaller to larger scales is met with in complicated fluid systems.

The meridional Ferrel cell is indirect in which ZKE is converted into ZAPE.

In the tropics, on the other hand, the main source for zonal kinetic energy is conversion from ZAPE by *y-p* overturnings in the Hadely cell (Pisharoty, 1954; Palmen, 1964; Kidson et al, 1969, Held I.M and Hou A.Y, 1980, Frierson D.M.W, Lu J, Chen G, 2007) (Fig. 2.1a).

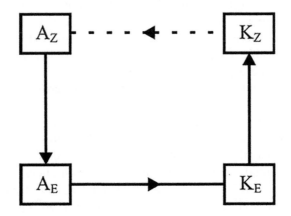

Fig. 2.1 Energy conversion in the extratropical atmosphere

Fig. 2.1a Energy conversion in the tropics

3. Atmospheric General Circulation

3.1 Introduction and General Considerations

The atmospheric wind system is driven by the sun. The sun's surface temperature is around 6000 °K. It emits radiation mainly in the visible, ultraviolet and infrared i.e., from 0.1 to 2 μ. The earth's atmosphere is more or less transparent to this radiation. Only ultraviolet radiation is absorbed by Ozone in the stratosphere. The earth's temperature is around 250 °K. The earth emits long wave radiation i.e., infrared radiation i.e., from 2 to 100 μ. The atmosphere – mainly H_2O, CO_2 etc., absorbs this radiation. The atmosphere acts like a greenhouse and keeps the earth warm. (Fig. 3.1).

The sun's radiation is not uniformly distributed throughout the latitudes because of the sun-earth geometry. The tropics receive more radiation than they lose. The higher latitudes lose more radiation than they receive (Fig. 3.2). Thus there is a positive balance or net heating in the tropics and negative balance or cooling in the higher latitudes. Thus, we are heating the atmosphere in the tropics and cooling it in the higher latitudes. There is a heat source in the tropics and a heat sink in the higher latitudes. We are differentially heating the fluid (atmospheric air). Thus we have a heat engine. This is a complex engine. It is sitting on a rotating platform i.e., the earth.

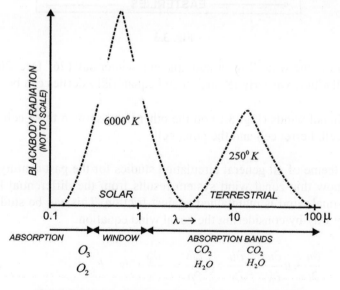

Fig. 3.1 Solar and Terrestrial radiation (Black body)

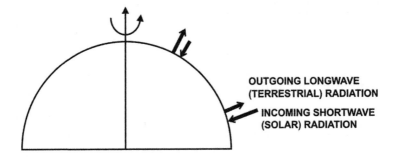

Fig. 3.2 Latitudinal distribution of radiation

The time average and sometimes zonally averaged wind system is often referred to as the general circulation. Fig. 3.3 shows the annually averaged, zonal wind circulation (schematic). It has strong westerly winds in the upper troposphere of middle latitudes. These are called westerly jet streams – strong winds from the west. The winds in the tropics are easterly (or from east).

Fig. 3.3

Fig. 3.4 shows the wind flow at 200 hpa on a winter day (26th Dec. 2018). The strong midlatitude westerly jet stream and equatorial esterlies can be seen.

The meridional winds (Fig. 3.5) on the other hand show a three cell structure – Hadley cell, Ferrel cell and the polar cell.

The main theme of all general circulation studies for the past century or more has been how this zonal wind system results from the differential heating – how this zonal wind is maintained against friction. This can be studied in an elementary way by considering the zonal wind equation.

$$\frac{\partial u}{\partial t} + u\frac{\partial u}{\partial x} + v\frac{\partial u}{\partial y} + \omega\frac{\partial \omega}{\partial p} = -\frac{\partial \phi}{\partial x} + fv + F_x \qquad \dots (3.1)$$

Add
$$u\left(\frac{\partial u}{\partial x}+\frac{\partial v}{\partial y}+\frac{\partial \omega}{\partial p}\right)=0 \qquad \qquad \dots (3.2)$$

$$\frac{\partial u}{\partial t}+\frac{\partial u^2}{\partial x}+\frac{\partial uv}{\partial y}+\frac{\partial u\omega}{\partial p}=-\frac{\partial \phi}{\partial x}+fv+F_x \qquad \dots (3.3)$$

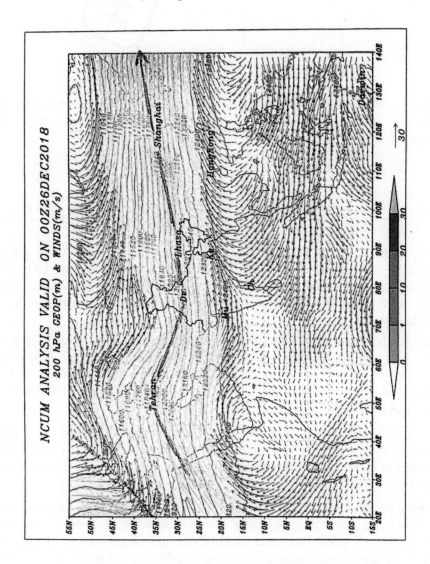

Fig. 3.4 Wind flow at 200 hPa, 26th Dec 2018

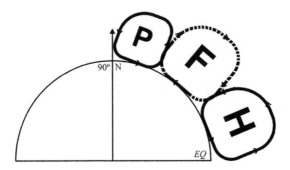

Fig. 3.5 Mean meridional circulation (schematic)

{The vertical scale of the meridional cells is highly exaggerated in the diagram. In the actual atmosphere these cells are confined to 10-20 km (troposphere)}.

Averaging over a latitude circle

$$\frac{\partial[u]}{\partial t} = -\frac{\partial[u][v]}{\partial y} - \frac{\partial\left[u^*v^*\right]}{\partial y} + f[v] + [F_x] + \text{mountain effect}^1 \quad \dots (3.4)$$

This is the equation for the maintenance of the mean zonal wind against frictional loss. It has been found that the largest terms are the eddy flux (second term) and the mean meridional flux (third term). In the middle latitudes the eddy flux is the largest. In the tropics the mean meridional flux is the largest.

Besides the mean zonal flow, there are stationary non-zonal flows. The most important is the Asian monsoon. In this the meridional circulation is reversed (southerlies in lower levels and northerlies in upper troposphere) (Fig. 3.6). The zonal winds are westerlies in the lower levels and easterly jet in the upper troposphere (Fig. 3.12b, Fig. 3.13b).

In the equatorial pacific there is an east-west vertical circulation i.e., Walker circulation – easterlies in the lower troposphere, upward motion over Indonesia, westerlies in the upper troposphere and sinking motion over eastern equatorial Pacific.

During northern hemispheric summer, the heating of the Asian continent with respect to the Indian ocean together with the secondary heat source caused by the release of latent heat over north India and neighborhood induces the monsoon meridional circulation with southerlies in the lower levels, rising motion over north India and neighborhood, northerlies in the upper troposphere and sinking motion over Indian ocean as indicated in Fig. 3.6.

---

[1]pressure gradient force resulting from east-west pressure difference across mountains

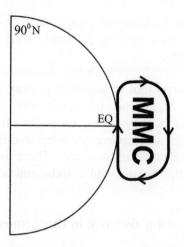

Fig. 3.6 Monsoon meridional circulation (schematic)

The southerlies in the lower levels turn right by coriolis force and result in monsoon westerlies. In the upper troposphere the northerlies turn right and result in the easterly jet. Fig. 3.14 shows the upper troposphere easterly jet (Koteswaram, 1958) on a summer monsoon day.

In nineteenth century and early twentieth century - in physics and other sister sciences progress occurred in two parallel branches like experiment and theory and interplay between these two branches. However in later twentieth century, in fields like atmospheric science, progress occurred in the following parallel branches:

- Observational studies
- Theoretical studies
- Laboratory simulation
- Computer simulation

Meteorologists collect observations of the atmosphere like pressure, temperature, wind, humidity etc., for weather forecasting. These observations are studied with a backdrop of theory – such as studies of momentum and heat transport, energy conversions etc. Phenomenonological synoptic studies of disturbances are also conducted. These are examples of observational studies. Using the basic equations of atmospheric science, theoretical studies like hydrodynamic instability of basic flows, growth of atmospheric disturbances, general circulation studies etc., are conducted. These are examples of theoretical studies.

In laboratory simulation experiments, water in a dish pan is differentially heated and it is placed on a rotating platform. Thus atmospheric flows are simulated in the laboratory.

However the most powerful tool of twentieth century is computer simulation using large computers. Right from the earlier days of small computers to present day supercomputers, they are used for simulation and also prediction. The basic equations of the atmosphere are integrated numerically to simulate general circulation, climate, monsoon etc. There have been attempts to simulate intraseasonal, interannual, and decadal and century and longer scale variability of climate.

3.2   Angular Momentum Balance in the Atmosphere

The angular momentum of unit mass of air at latitude ϕ is

$$M = (u + \Omega r) r$$

where

a = earth's radius

$r = a \cos \phi$ (Fig. 3.7)

$M = ur + \Omega r^2$

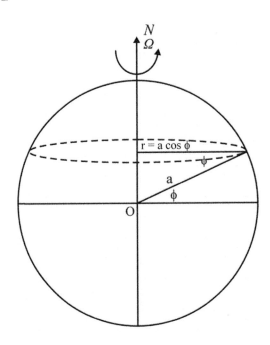

Fig. 3.7

M is absolute angular momentum. The first term on the right is the relative angular momentum and the second term is due to earth's rotation.

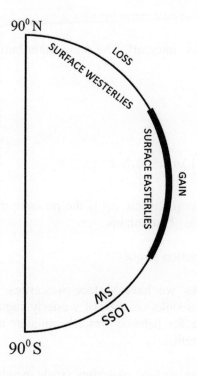

Fig. 3.8 Mean zonal surface winds (schematic)

The absolute angular momentum M about the earth's axis can change only as a result of torques due to pressure gradient and frictional forces.

So the equation for the rate of change of angular momentum is

$$\frac{dM}{dt} = \left(-\alpha \frac{\partial p}{\partial x} + F_x \right) a \cos \phi$$

Integrating this over the entire atmospheric volume poleward of a latitude,

$$\int \frac{\partial}{\partial t} (\rho M) d\tau = -\int \nabla \cdot (\rho M \vec{V}) d\tau - \int \frac{\partial}{\partial x} (p a \cos \phi) d\tau + \int \rho F_x \, a \cos \phi \, d\tau$$

where $d\tau = dx \, dy \, dz$

The first term (on the right hand side) is the angular momentum transport

$$= \int \rho M V_n \, d\sigma$$

(transforming to surface integral). The main contribution is at the southern wall.

$V_n = 0$ at the surface

$\rho = 0$ at the top.

The second term $= -\iint \Delta p \, a \cos \phi \, dy \, dz$

This is the mountain torque where Δp is the pressure difference between the eastern and western side of mountains.

The third term is the friction torque.

In the middle latitudes we have surface westerlies. Surface friction and mountain torques act as sinks i.e., loss of westerly angular momentum of the atmosphere. Therefore for balance westerly angular momentum has to be transported from the tropics.

In the tropics there are surface easterlies (trade winds). Atmosphere gains westerly angular momentum in the tropics and this is transported polewards to middle latitudes. (Fig. 3.8).

The total poleward transport of angular momentum through a latitudinal wall is

$$\frac{2\pi a^2 \cos^2 \phi}{g} \int_0^{p_0} \left\{ \Omega[v] a \cos \phi + [u][v] + [u^* v^*] \right\} dp$$

The first term is called the omega transport term. The second term is the drift term and the third term is the eddy transport term. In the first term

$$\int_0^{p_0} [v] \, dp = 0$$

as there is no net mass transport across a latitude.

Therefore, omega transport term goes to zero.

Observational studies indicate that in the middle latitudes the main poleward transport is accomplished by eddies or large scale (Rossby) waves. These waves are titled NNE to SSW.

In the tropics including the monsoon region on the other hand, the eddy transport term is small and the main contribution is from the omega transport term (mean meridional circulation).

3.3   Three Dimensional Structure of the General Circulation

Let us now look at the three dimensional structure of the observed general circulation of the atmosphere. Recently NCEP and European Centre have prepared excellent reanalysis products. This should be considered a great boon for the meteorological community. I have obtained the following maps from Dr. Ayantika Dey Choudhary of CCCR, IITM based on the above reanalysis. This is gratefully acknowledged.

Fig. 3.9 (a, b) show the vertically integrated diabatic heating fields during northern winter (Jan) and northern summer (July) respectively. These show some very interesting features.

During the northern winter (Jan, Fig. 3.9a) the near equatorial heating inducing the rising branch of the Hadley cell is not uniform throughout the globe. Notice that the maximum heating is over the equatorial western Pacific, Indonesia, tropical South America and tropical Africa (southern hemisphere). The extra tropical cooling is concentrated over the extra tropical and polar regions of Eurasia and North America. Notice also the cooler regions over eastern equatorial Pacific and Atlantic (and adjoining southern tropical oceanic regions).

During the northern summer (July, Fig. 3.9b), on the other hand, the maximum heating regions are mainly over the south Asian monsoon region, equatorial western Pacific and equatorial (central) Americas. The heating over the monsoon region is contributed by the release of latent heat in the rising upward branch of the monsoon meridional (or reverse Hadley as it is sometimes referred to) cell. The heating of the Eurasian and North American continents is also evident. Notice the cold regions over eastern Pacific, eastern Atlantic and of course the polar regions, noticeably in the winter hemisphere.

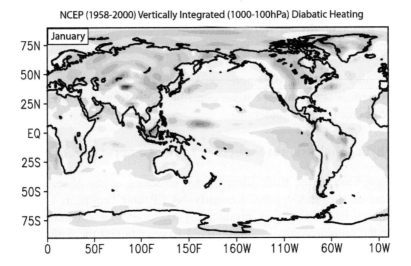

Fig. 3.9a Vertically Integrated Diabatic Heating Field, January

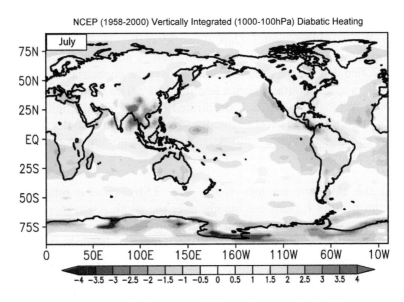

Fig. 3.9b Vertically Integrated Diabatic Heating Field, July

These diabatic heating fields or rather their gradients induce divergent wind (velocity potential) circulations as shown in Fig. 3.10 (a, b) and Fig. 3.11 (a, b).

During northern winter at 850 hpa (Fig. 3.10a), there is marked inflow into the convergent region over equatorial western Pacific and to a lesser extent over equatorial Indian ocean and tropical South America. Thus, the rising branch of the Hadley circulation is most pronounced over equatorial western pacific and equatorial Indian ocean. The sinking branch (low level divergent region), on the other hand is seen over Sahara desert, middle eastern region and over subtropical Atlantic (north) and neighbourhood.

Now if we look at the upper troposphere (200 hpa, Fig. 3.11a) we see a complimentary picture of pronounced outflow, as part of the upper branch of the Hadley cell, over equatorial western Pacific and adjoining Indian ocean. Upper convergence and thus sinking regions are seen over Sahara, middle east and subtropical north Atlantic.

Thus the Hadley cell is most pronounced over western pacific and adjoining south Asia, Africa and to a lesser extent over tropical south America.

It is now time to look at the northern summer (July) velocity potential (divergent wind) field in the upper troposphere (Fig. 3.11b). Notice the pronounced outflow over the south Asian monsoon region-tropical western Pacific, southeast Asia, India and China. Notice pronounced northerlies over the monsoon region as part of the upper branch of the monsoon meridional cell. The convergent (sinking motion) regions, on the other hand are seen over subtropical regions of (south) Indian ocean, southern Africa and south Atlantic. This is the descending branch of the monsoon meridional cell.

In the lower troposphere (Fig. 3.10b), we see regions of strong convergence over the monsoon region extending from north India to tropical western pacific. Strong southerly meridional winds are seen over Arabian sea, Bay of Bengal and extend up to east of Indonesia. Regions of divergence are seen over tropical south Atlantic and adjoining southern Africa and south America.

These divergent wind circulations induced by the diabatic heating gradients result in the observed wind field under the influence of earth's rotation (Coriolis force). (Fig. 3.12 (a, b)), and (Fig. 3.13 (a, b)). In addition to the effect of diabatic heating gradients, there are also other effects like topography and the role of eddies (as discussed in chapter 2) etc. In particular, during northern summer, the Tibetan plateau acts as an elevated heat source. The orographic rainfall (and associated release of latent heat) over northeast India, induced by the steep topographic barrier, is an additional heat source. The Himalayas also help to confine the lower tropospheric moist monsoon winds to the south of the Himalayas. Even the east African topography has been shown to be important for monsoon circulation.

In the Hadley cell, during winter, the northerlies in the lower levels turn right and result in the northeast trades in the northern hemisphere.

During the northern summer, on the other hand, the southerlies of the lower branch of the monsoon meridional cell result in the monsoon southwesterlies and westerlies (by Coriolis turning) over India and southeast Asia.

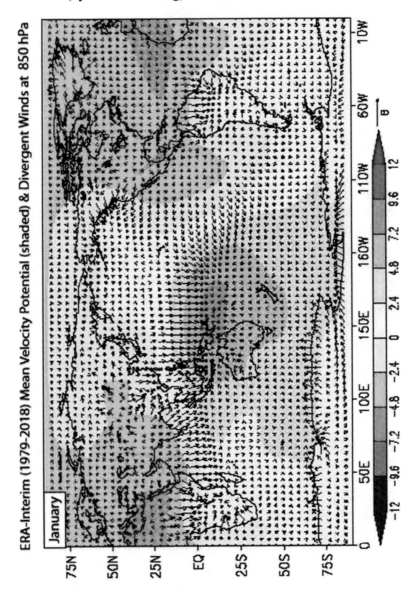

Fig. 3.10a Mean velocity potential (divergent wind) field, 850 hpa, January

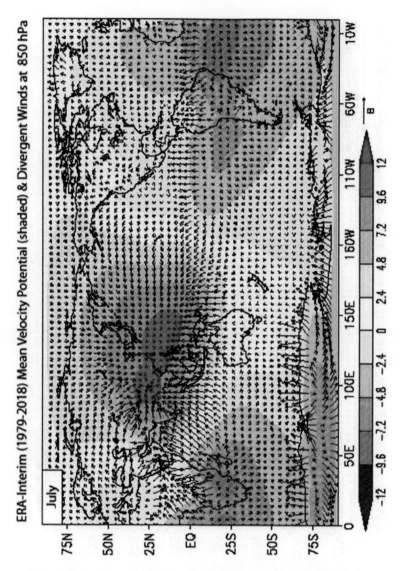

Fig. 3.10b Mean velocity potential (divergent wind) field, 850 hpa, July

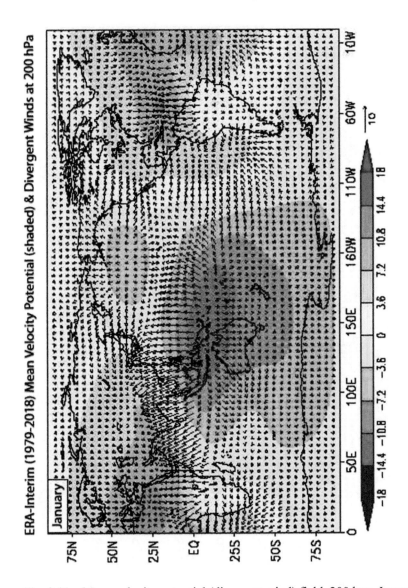

Fig. 3.11a Mean velocity potential (divergent wind) field, 200 hpa, January

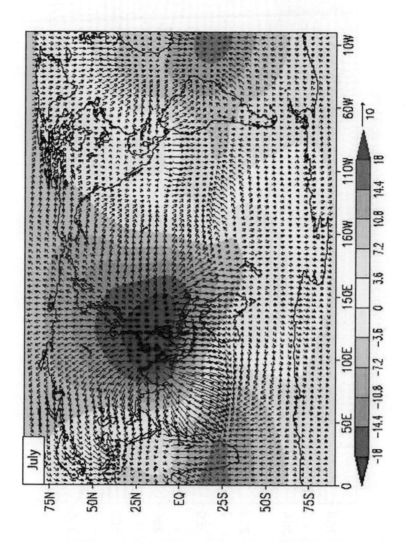

Fig. 3.11b Mean velocity potential (divergent wind) field, 200 hpa, July

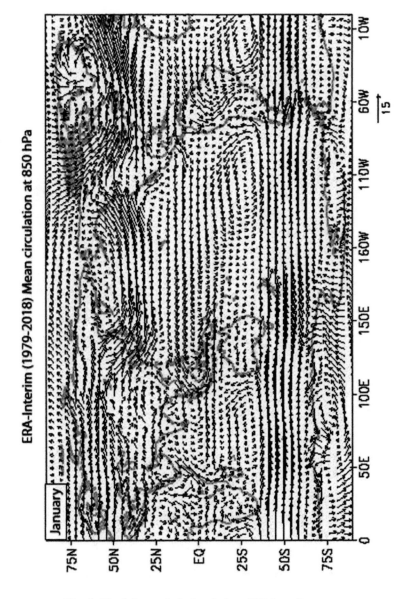

Fig. 3.12a Mean wind circulation, 850 hpa, January

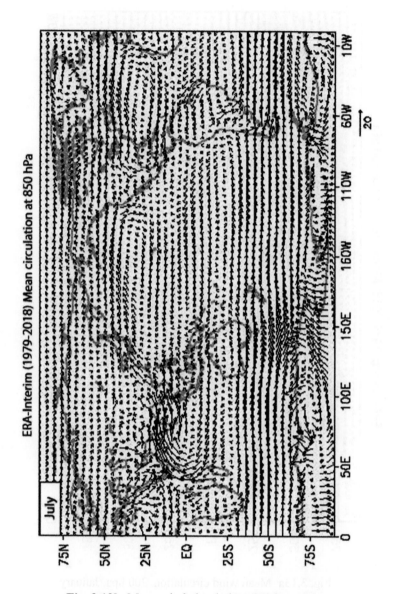

Fig. 3.12b Mean wind circulation, 850 hpa, July

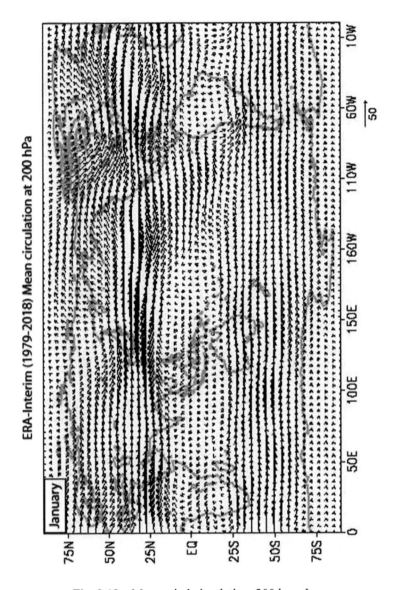

Fig. 3.13a Mean wind circulation, 200 hpa, January

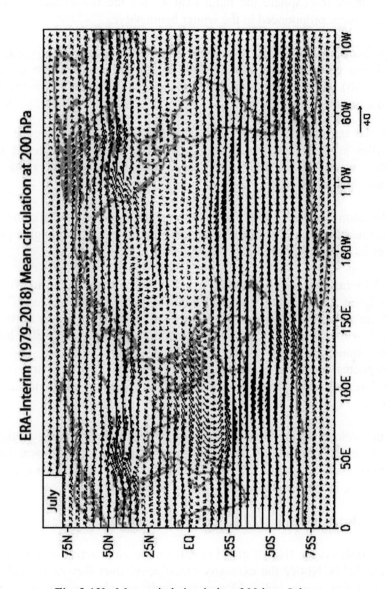

Fig. 3.13b Mean wind circulation, 200 hpa, July

In the upper troposphere the main features are the midlatitude westerly jet stream – more pronounced in the winter hemisphere.

In the monsoon region the northerlies i.e., the upper branch of the monsoon meridional cell turn right (by Coriolis force) and result in the easterly jet (Fig. 3.13b). The Tibetan anticyclone is also seen. Fig. 3.14 shows the easterly jet on a summer monsoon day.

Over the equatorial Pacific notice the east-west Walker circulation with easterlies in the lower troposphere, rising motion over equatorial western Pacific, westerlies in the upper troposphere and sinking motion over equatorial eastern Pacific.

In the monsoon region (northern summer) the large diabatic heating is contributed by release of latent heat. Thus the monsoon heat engine produces part of its fuel while running.

Fig. 3.15 (a, b, c, d, e) show the wind flow maps (taken from NCMRWF analysis) at 925 hpa, 850 hpa, 700 hpa, 500 hpa and 200 hpa on 6th August 2019, an active monsoon day. Notice the low pressure area over north Bay of Bengal with associated cyclonic circulation extending upto 500 hpa. The strong monsoon southwesterlies / westerlies in the lower and middle tropospheres and the monsoon easterly jet in the upper troposphere can be seen. The east-west shear zone in the middle troposphere over north peninsula is a typical feature of active monsoon.

An anticyclone in the upper troposphere is seen over Pakistan, Afghanistan, Iran and neighbourhood. In the mean, an anticyclone is located over Tibet and is known as the Tibetan high or Tibetan anticyclone. A westward shift of the anticyclone is often noticed during active monsoon epochs.

We can also notice tropical disturbances over tropical northwest Pacific.

Fig. 3.16 shows the INSAT cloud picture on the same active monsoon day (6th August 2019). Notice the extensive cloud cover over Bay of Bengal, central and adjoining north India. Monsoon clouds are also seen over peninsular India.

Cloud associated with the tropical disturbances over northwest pacific can also be seen.

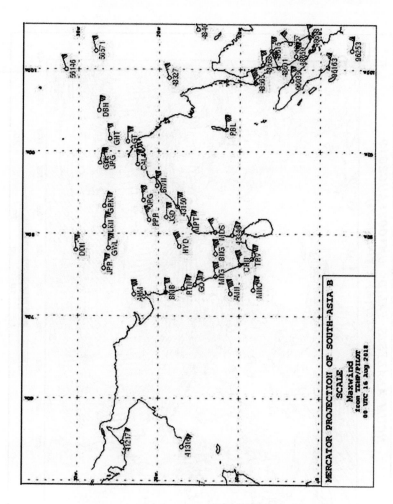

Fig. 3.14 Easterly jet stream 16th Aug 2018

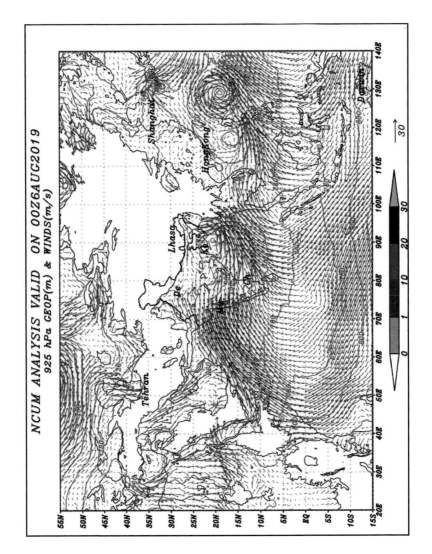

Fig. 3.15a Wind flow at 925 hpa, 6th Aug 2019

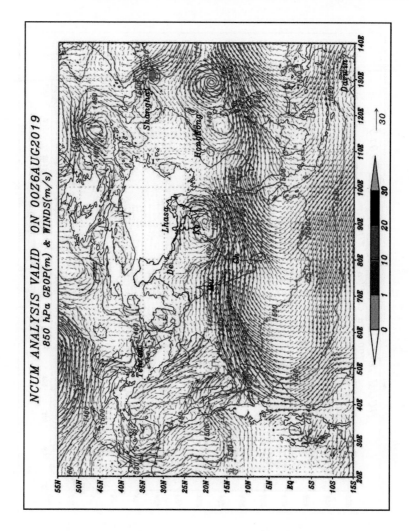

Fig. 3.15b Wind flow at 850 hpa, 6th Aug 2019

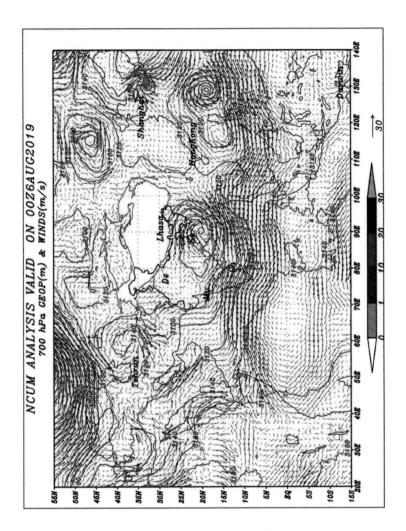

Fig. 3.15c Wind flow at 700 hpa, 6th Aug 2019

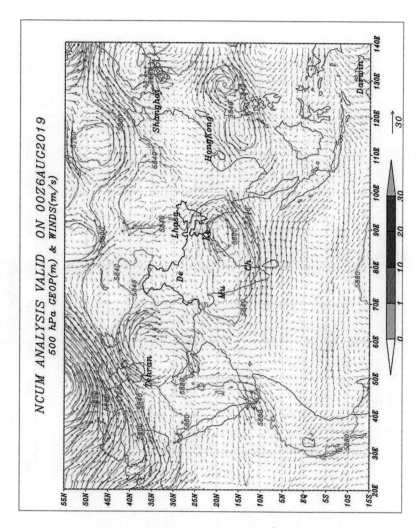

Fig. 3.15d Wind flow at 500 hpa, 6th Aug 2019

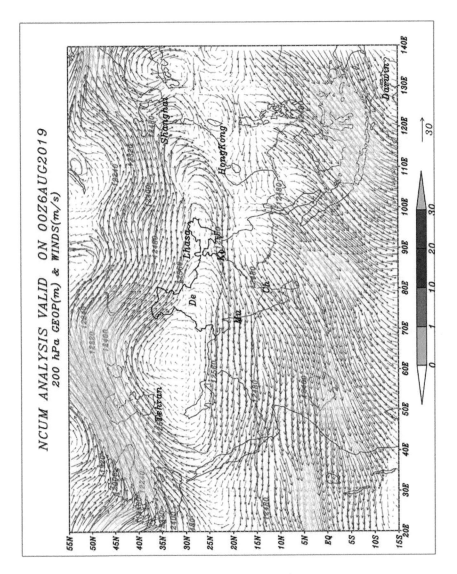

Fig. 3.15e Wind flow at 200 hpa, 6th Aug 2019

SAT :INSAT-3D IMG   06-08-2019/(1500 to 1527) GMT
IMG_TIR1 10.8 um   06-08-2019/(2030 to 2057) IST
L1B FULL DISK (LINEAR STRETCH: 1.0%)

IMD/Delhi

476   909

Fig. 3.16 INSAT cloud picture, 6th Aug 2019

4. Waves in the Atmosphere

4.1 Introduction

The waves we commonly come across in the atmosphere are:

   (i)   Sound waves
   (ii) Gravity waves
   (iii) Rossby waves
   (iv) Kelvin waves

Sound waves

Sound waves are longitudinal waves consisting of alternating regions of compression and rarefaction. The particle oscillations are along the direction of propagation.

The speed of propagation of sound is

$$c = u \pm (\gamma RT)^{1/2}$$

where $\gamma = \dfrac{C_P}{C_V}$

Gravity waves

Shallow water gravity waves are transverse waves. The restoring force is gravity which is in the vertical direction whereas the direction of propagation is horizontal. These waves travel eastwards or westwards with phase speed.

$$c = u \pm \sqrt{gH}$$

where H is the depth of the fluid and is small compared to the wavelength.

In the atmosphere there are also internal gravity waves.

Sound and gravity waves are however, not very important for weather and climate.

Rossby waves

The waves that are important for weather and climate are Rossby waves. The rotation of the earth or to be more precise, the variation of Coriolis parameter with latitude i.e., $\beta = \dfrac{df}{dy}$ is important for these waves.

The phase speed of Rossby waves is

$$c = u - \frac{\beta}{k^2}$$

where the wave number $k = \dfrac{2\pi}{L}$, where L is the wave length.

These waves move westwards with respect to the basic flow u. Shorter Rossby waves usually travel eastwards with respect to the earth. Typical wavelength in the middle latitudes is of the order 6000 km. Typical phase speeds are of the order 10 ms^{-1}. These are dispersive waves i.e., the phase speed depends upon the wavelength. Rotational motion dominates.

We can derive an expression for the speed of Rossby waves.

4.2   Speed of Rossby Waves

Consider the non-divergent barotropic vorticity equation

$$\frac{d\zeta_a}{dt} = 0 , \zeta_a = \zeta + f$$

$$\frac{d}{dt}(\zeta + f) = 0$$

$$\frac{\partial \zeta}{\partial t} + \vec{V} \cdot \nabla(\zeta + f) = 0$$

$$\frac{\partial \zeta}{\partial t} + u \frac{\partial \zeta}{\partial x} + v \frac{\partial \zeta}{\partial y} + \beta v = 0$$

Consider a basic flow $U = $ constant, $\overline{v} = 0$, $\overline{\zeta} = 0$

$$U = U + u' , \quad v = v' , \quad \zeta = \zeta'$$

The perturbation vorticity equation is

$$\frac{\partial \zeta'}{\partial t} + U \frac{\partial \zeta'}{\partial x} + u' \frac{\partial \zeta'}{\partial x} + v' \frac{\partial \zeta'}{\partial y} + \beta v' = 0$$

Neglecting the second order terms

$$\frac{\partial \zeta'}{\partial t} + U \frac{\partial \zeta'}{\partial x} + \beta v' = 0$$

$$\left(\frac{\partial}{\partial t} + U \frac{\partial}{\partial x} \right) \zeta' + \beta v' = 0$$

$$\zeta' = \nabla^2 \psi', \, v' = \frac{\partial \psi'}{\partial x}$$

$$\left(\frac{\partial}{\partial t} + U \frac{\partial}{\partial x} \right) \nabla^2 \psi' + \beta \frac{\partial \psi'}{\partial x} = 0$$

Assuming wave solutions of the type

$$\psi' = \psi \, e^{ik(x-ct)}$$

$$v' = \frac{\partial \psi'}{\partial x} = ik \psi'$$

$$\nabla^2 \psi' = \frac{\partial^2 \psi'}{\partial x^2} + \frac{\partial^2 \psi'}{\partial y^2} = -k^2 \psi'$$

$$\frac{\partial}{\partial t} \sim -ikc$$

Substituting, we obtain

$$ik(U-c)\left(-k^2 \psi'\right) + ik\beta \psi' = 0$$

which yields $U - c = \dfrac{\beta}{k^2}$

or the phase speed of Rossby waves

$$c = U - \frac{\beta}{k^2} = U - \frac{\beta L^2}{4\pi^2}$$

where $k = \dfrac{2\pi}{L}$

Rossby waves move westwards with respect to the basic flow. They are dispersive waves i.e., the phase speed is a function of the wavelength.

Figures 4.1 and 4.2 show midlatitude Rossby waves at 500 hpa and 200 hpa on a winter day (27th Dec. 2018).

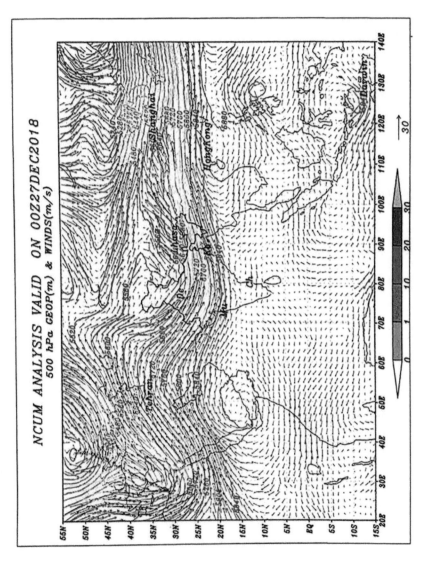

Fig. 4.1 Wind flow at 500 hpa, 27th Dec 2018

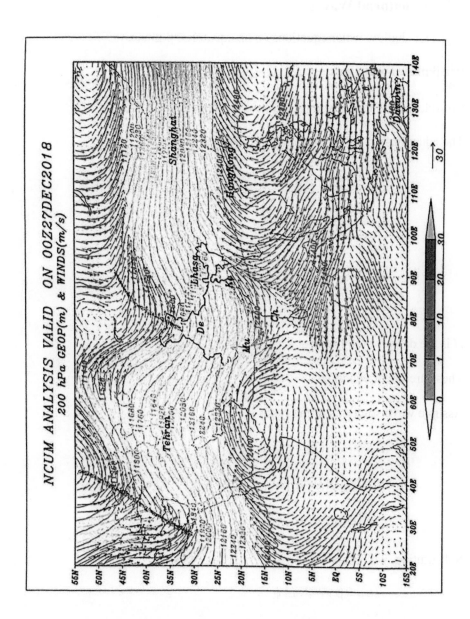

Fig. 4.2 Wind flow at 200 hpa, 27th Dec 2018

4.3   Equatorial Waves

Consider shallow water equations on an equatorial β plane.

In a β plane approximation $f = f_0 + \beta y = \beta y$

Consider small perturbations on a zero basic state.

The equations are (after Matsuno, 1966)

$$\frac{\partial u'}{\partial t} = -\frac{\partial \phi'}{\partial x} + \beta y v'$$

$$\frac{\partial v'}{\partial t} = -\frac{\partial \phi'}{\partial y} - \beta y u'$$

$$\frac{\partial \phi'}{\partial t} + g H \left(\frac{\partial u'}{\partial x} + \frac{\partial v'}{\partial y} \right) = 0$$

H is the depth of the fluid.

These are linear equations.

Assume wave solutions of the type

$$u' = u \, \exp i(kx - \sigma t)$$

$$v' = v \exp i(kx - \sigma t)$$

$$\phi' = \phi \exp i(kx - \sigma t)$$

Substituting, we obtain

$$-i\sigma u - \beta y v = -ik\phi$$

$$-i\sigma v + \beta y u = -\frac{\partial \phi}{\partial y}$$

$$-i\sigma \phi + gH \left(iku + \frac{\partial v}{\partial y} \right) = 0$$

Eliminating *u* and φ,

we obtain an equation in v

$$\frac{d^2v}{dy^2} + \left[\left(\frac{\sigma^2}{gH} - k^2 - \frac{k}{\sigma}\beta \right) - \frac{\beta^2}{gH}y^2 \right]v = 0$$

This is a second order ordinary differential equation. This is formally similar to Shroedinger's Harmonic oscillator equation.

As $y \to \infty$, v is bounded i.e., decays at large y. Under this (boundary) condition there are only eigen (particular) solutions when

$$\frac{\sqrt{gH}}{\beta} \left(-\frac{k}{\sigma}\beta - k^2 + \frac{\sigma^2}{gH} \right) = 2n+1$$

$n = 0, 1, 2,$ etc.

For the above eigen values, the eigen solutions are

$$v \sim e^{-\frac{1}{2}\xi^2} H_n(\xi)$$

where $\qquad \xi = \left(\frac{\beta}{\sqrt{gH}} \right)^{1/2} y$

H_n are Hermite polynomials.

These solutions are westward and eastward moving equatorial waves. Equator acts like a wave guide. These waves are equatorially trapped.

We obtain westward moving Rossy waves, eastward and westward moving gravity waves, as also mixed Rossby-gravity waves.

If we consider the special case of $v = 0$, we obtain eastward moving Kelvin waves.

Fig. (4.3) shows wind flow pattern at 850 hpa on 11th December 2018 taken from NCMRWF analysis. Observe the equatorial wave over equatorial Indian Ocean and adjoining Bay of Bengal. This is apparently an equatorial Rossby wave (Gill, 1980; Keshavamurty, 1982). In this case it is modified by latent heating. Kasture et.al. (1991) showed, using linear analytical calculations and nonlinear numerical studies, that inclusion of moisture reduces the speed of equatorial Kelvin as well as of Rossby waves.

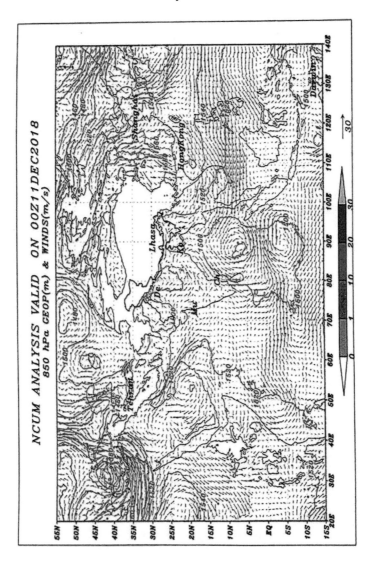

Fig. 4.3 Wind flow at 850 hpa, 11th December, 2018

5. Large Scale Atmospheric Modeling

5.1 Introduction

The basic governing equations of atmospheric flow are

$$\frac{du}{dt} = -\alpha\frac{\partial p}{\partial x} + f\,v + F_x$$

$$\frac{dv}{dt} = -\alpha\frac{\partial p}{\partial y} - f\,u + F_y$$

$$\frac{dw}{dt} = -\alpha\frac{\partial p}{\partial z} - g + F_z$$

$$\frac{1}{\rho}\frac{d\rho}{dt} = -\nabla\cdot\vec{V}$$

$$p\alpha = RT$$

$$C_v\frac{dT}{dt} + p\frac{d\alpha}{dt} = C_p\frac{dT}{dt} - \alpha\frac{dp}{dt} = \dot{Q}$$

The first three equations are equations of motion (Newton's second law of motion) along the *x*, *y* and *z* directions. The fourth edquation is continuity equation (law of conservation of mass). The fifth equation is the equation of state (combined Charles and Boyle's law). The sixth equation is the first law of thermodynamics.

Thus we have six equations in six unknowns, i.e., pressure *p*, specific volume α (or ρ density), temperature *T* and the three components of velocity *u*, *v* and *w*. These equations are first order nonlinear partial differential equations and in general cannot be solved analytically. Thus we solve them numerically using large computers. The above equations are known as primitive (complete) equations. In numerical weather prediction models and climate models these equations are used. Knowing the present state of the atmosphere we can predict the future state using the above equations. (There are, of course some limitations).

However, for research purposes and theoretical or quasi-theoretical studies simplified model formulations are used. These are discussed in the following sections.

5.2   Vorticity Equation

It is convenient to get an equation for the time rate of change of the vertical component of vorticity ζ. We start with the equations of motion along the x and y directions, in x, y, p co-ordinates.

$$\frac{du}{dt} = -\frac{\partial \phi}{\partial x} + fv + F_x$$

$$\frac{dv}{dt} = -\frac{\partial \phi}{\partial y} - fu + F_y$$

Expanding,

$$\frac{\partial u}{\partial t} + u\frac{\partial u}{\partial x} + v\frac{\partial u}{\partial y} + \omega\frac{\partial u}{\partial p} = -\frac{\partial \phi}{\partial x} + fv + F_x \qquad \dots (5.1)$$

$$\frac{\partial v}{\partial t} + u\frac{\partial v}{\partial x} + v\frac{\partial v}{\partial y} + \omega\frac{\partial v}{\partial p} = -\frac{\partial \phi}{\partial y} - fu + F_y \qquad \dots (5.2)$$

We take $\dfrac{\partial}{\partial y}$ (5.1)

$$\frac{\partial}{\partial t}\left(\frac{\partial u}{\partial y}\right) + \frac{\partial u}{\partial y}\frac{\partial u}{\partial x} + u\frac{\partial}{\partial x}\left(\frac{\partial u}{\partial y}\right) + \frac{\partial v}{\partial y}\frac{\partial u}{\partial y} + v\frac{\partial}{\partial y}\left(\frac{\partial u}{\partial y}\right) +$$

$$\frac{\partial \omega}{\partial y}\frac{\partial u}{\partial p} + \omega\frac{\partial}{\partial p}\left(\frac{\partial u}{\partial y}\right) = -\frac{\partial^2 \phi}{\partial y\,\partial x} + f\frac{\partial v}{\partial y} + v\frac{\partial f}{\partial y} + \frac{\partial F_x}{\partial y} \qquad \dots (5.3)$$

and $\dfrac{\partial}{\partial x}$ (5.2)

$$\frac{\partial}{\partial t}\left(\frac{\partial v}{\partial x}\right) + \frac{\partial u}{\partial x}\frac{\partial v}{\partial x} + u\frac{\partial}{\partial x}\left(\frac{\partial v}{\partial x}\right) + \frac{\partial v}{\partial x}\frac{\partial v}{\partial y} + v\frac{\partial}{\partial y}\left(\frac{\partial v}{\partial x}\right) +$$

$$\frac{\partial \omega}{\partial x}\frac{\partial v}{\partial p} + \omega\frac{\partial}{\partial p}\left(\frac{\partial v}{\partial x}\right) = -\frac{\partial^2 \phi}{\partial x\,\partial y} - f\frac{\partial u}{\partial x} + \frac{\partial F_y}{\partial x} \qquad \dots (5.4)$$

Subtracting equation (5.3) from equation (5.4) yields

$$\frac{\partial}{\partial t}\left(\frac{\partial v}{\partial x}-\frac{\partial u}{\partial y}\right)+u\frac{\partial}{\partial x}\left(\frac{\partial v}{\partial x}-\frac{\partial u}{\partial y}\right)+v\frac{\partial}{\partial y}\left(\frac{\partial v}{\partial x}-\frac{\partial u}{\partial y}\right)+$$

$$\omega\frac{\partial}{\partial p}\left(\frac{\partial v}{\partial x}-\frac{\partial u}{\partial y}\right)+\frac{\partial u}{\partial x}\left(\frac{\partial v}{\partial x}-\frac{\partial u}{\partial y}\right)+\frac{\partial v}{\partial y}\left(\frac{\partial v}{\partial x}-\frac{\partial u}{\partial y}\right)+$$

$$\frac{\partial\omega}{\partial x}\frac{\partial v}{\partial p}-\frac{\partial\omega}{\partial y}\frac{\partial u}{\partial p}=-f\left(\frac{\partial u}{\partial x}+\frac{\partial v}{\partial y}\right)-\beta v+\frac{\partial F_y}{\partial x}-\frac{\partial F_x}{\partial y}$$

Simplifying, we get

$$\frac{\partial\zeta}{\partial x}+u\frac{\partial\zeta}{\partial x}+v\frac{\partial\zeta}{\partial y}+\omega\frac{\partial\zeta}{\partial p}+\zeta D+\frac{\partial\omega}{\partial x}\frac{\partial v}{\partial p}-\frac{\partial\omega}{\partial y}\frac{\partial u}{\partial p}=$$

$$-fD-\beta v+\frac{\partial F_y}{\partial x}-\frac{\partial F_x}{\partial y}$$

or $\qquad \dfrac{d\zeta}{dt}=-(\zeta+f)D-\beta v+\dfrac{\partial\omega}{\partial y}\dfrac{\partial u}{\partial p}-\dfrac{\partial\omega}{\partial x}\dfrac{\partial v}{\partial p}+\dfrac{\partial F_y}{\partial x}-\dfrac{\partial F_x}{\partial y}$

Now, $\qquad \dfrac{df}{dt}=\dfrac{\partial f}{\partial t}+u\dfrac{\partial f}{\partial x}+v\dfrac{\partial f}{\partial y}+\omega\dfrac{\partial f}{\partial p}=\beta v$

$\beta=\dfrac{df}{dy}$ is known as the Rossy parameter.

So, $\qquad \dfrac{d(\zeta+f)}{dt}=-(\zeta+f)D-\hat{k}\cdot\nabla\omega\times\dfrac{\partial\vec{V}}{\partial p}+\left(\dfrac{\partial F_y}{\partial x}-\dfrac{\partial F_x}{\partial y}\right)$

This is the vorticity equation. The first term on the right hand side is known as the divergence term and is the largest for large scale motion. The second term is the twisting or tilting term and the third is the friction term.

As an approximation

$$\frac{d\zeta_a}{dt}=-\zeta_a D\,,$$

where $\zeta_a=\zeta+f$ is the absolute vorticity

ζ is the relative vorticity.

5.3   Divergence Equation

To derive the divergence equation take

$$\frac{\partial}{\partial x} \text{ of (5.1) and } \frac{\partial}{\partial y} \text{ of (5.2) and add}$$

$$\frac{\partial}{\partial t}\left(\frac{\partial u}{\partial x}\right) + \frac{\partial u}{\partial x}\frac{\partial u}{\partial x} + u\frac{\partial}{\partial x}\left(\frac{\partial u}{\partial x}\right) + \frac{\partial v}{\partial x}\frac{\partial u}{\partial y} + v\frac{\partial}{\partial y}\left(\frac{\partial u}{\partial x}\right) +$$

$$\frac{\partial \omega}{\partial x}\frac{\partial u}{\partial p} + \omega\frac{\partial}{\partial p}\left(\frac{\partial u}{\partial x}\right) = -\frac{\partial^2 \phi}{\partial x^2} + f\frac{\partial v}{\partial x}$$

$$\frac{\partial}{\partial t}\left(\frac{\partial v}{\partial y}\right) + \frac{\partial u}{\partial y}\frac{\partial v}{\partial x} + u\frac{\partial}{\partial x}\left(\frac{\partial v}{\partial y}\right) + \frac{\partial v}{\partial y}\frac{\partial v}{\partial y} + v\frac{\partial}{\partial y}\left(\frac{\partial v}{\partial y}\right) +$$

$$\frac{\partial \omega}{\partial y}\frac{\partial v}{\partial p} + \omega\frac{\partial}{\partial p}\left(\frac{\partial v}{\partial y}\right) = -\frac{\partial^2 \phi}{\partial y^2} - f\frac{\partial u}{\partial y} - \beta u$$

i.e.,
$$\frac{\partial D}{\partial t} + u\frac{\partial D}{\partial x} + v\frac{\partial D}{\partial y} + \omega\frac{\partial D}{\partial p} + \left(\frac{\partial u}{\partial x}\right)^2 + \left(\frac{\partial v}{\partial y}\right)^2 +$$

$$2\frac{\partial v}{\partial x}\frac{\partial u}{\partial y} + \frac{\partial \omega}{\partial x}\frac{\partial u}{\partial p} + \frac{\partial \omega}{\partial y}\frac{\partial v}{\partial p} = -\nabla^2\phi + f\left(\frac{\partial v}{\partial x} - \frac{\partial u}{\partial y}\right) - \beta u$$

$$\frac{dD}{dt} + D^2 - 2\frac{\partial u}{\partial x}\frac{\partial v}{\partial y} + 2\frac{\partial v}{\partial x}\frac{\partial u}{\partial y} + \frac{\partial \omega}{\partial x}\frac{\partial u}{\partial p} + \frac{\partial \omega}{\partial y}\frac{\partial v}{\partial p}$$

$$= -\nabla^2\phi + f\zeta - \beta u$$

$$\frac{dD}{dt} + D^2 + 2J(v, u) + \nabla\omega \cdot \frac{\partial \vec{V}}{\partial p} = -\nabla^2\phi + f\zeta - \beta u$$

If we omit the smaller order terms, we obtain

$$\nabla^2\phi - f\zeta + \beta u + 2J(v, u) = 0$$

This is known as nonlinear balance equation.

If we also take into account energetic considerations, we obtain

$$\nabla^2\phi - f\zeta + \beta u_\psi + 2J(v_\psi, u_\psi) = 0$$

Simplifying further, we get

$$\nabla^2\phi - f\zeta + \beta u_\psi = 0$$

This is the linear balance equation.

Further simplifying,

$$\nabla^2\phi - f_0\zeta = 0$$

This is geostrophic relation (constant f).

If we retain terms of similar orders of magnitude in the vorticity and thermodynamic energy equations and use continuity equation and equation of state for elimination of variables then we obtain a hierarchy of models.

5.4   Hierarchy of Models

(i)   Non divergent barotropic Model
(ii)   Quasi – geostrophic Model
(iii)   Linear Balance Model
(iv)   Non linear Balance Model
(v)   Primitive equation Model

Filtered Models

5.4.1 Non Divergent Barotropic Model

$$\frac{d\zeta_a}{dt} = -\zeta_a \nabla \cdot \vec{V}$$

Here, divergence = 0

$$\frac{d\zeta_a}{dt} = 0$$

ζ_a, the absolute vorticity is conserved

$$\zeta = \nabla^2\psi = \frac{1}{f_0}\nabla^2\phi$$

$$\frac{\partial\zeta}{\partial t} + V_\psi \cdot \nabla(\zeta + f) = 0$$

$$\rightarrow \frac{\partial}{\partial t}(\nabla^2\psi) + V_\psi \cdot \nabla(\zeta + f) = 0$$

This is a prediction equation. Knowing the present ψ field we can predict the future field using this equation. This is the principle of numerical weather prediction (NWP).

5.4.2 Quasi-geostrophic Model

This model uses both the vorticity equation and the thermodynamic energy equation. This model is widely used for theoretical studies.

$$\frac{\partial}{\partial t}\left(\frac{1}{f_0}\nabla^2\phi\right)+\frac{1}{f_0}k\times\nabla\phi\cdot\nabla\left(\frac{1}{f_0}\nabla^2\phi+f\right)=f_0\frac{\partial\omega}{\partial p} \qquad \dots (5.5)$$

$$\frac{\partial}{\partial t}\left(\frac{\partial\phi}{\partial p}\right)+\vec{V}_g\cdot\nabla\left(\frac{\partial\phi}{\partial p}\right)+\sigma\omega=-\frac{R}{C_p p}\dot{Q} \qquad \dots (5.6)$$

Thus we have two equation in 2 unknowns ϕ and ω.

In the following, we derive the thermodynamic energy equation in the appropriate form

$$C_P\frac{dT}{dt}-\alpha\frac{dp}{dt}=\dot{Q}$$

Dividing by T, we get

$$C_p\frac{d\ln T}{dt}-R\frac{d\ln p}{dt}=\frac{\dot{Q}}{T}$$

It is convenient to use the potential temperature

$$\theta=T\left(\frac{p_0}{p}\right)^\kappa$$

Taking logarithmic derivative

$$\frac{d\ln\theta}{dt}=\frac{d\ln T}{dt}+\kappa\frac{d\ln p_0}{dt}-\kappa\frac{d\ln p}{dt}$$

Multiplying by C_p

$$C_p\frac{d\ln\theta}{dt}=C_p\frac{d\ln T}{dt}-R\frac{d\ln p}{dt}$$

Thus, $C_p \dfrac{d \ln \theta}{dt} = \dfrac{\dot{Q}}{T}$

$$\frac{d \ln \theta}{dt} = \frac{\dot{Q}}{C_p T}$$

$$\frac{1}{\theta} \frac{d\theta}{dt} = \frac{1}{C_p T} \dot{Q}$$

$$\frac{d\theta}{dt} = \frac{\theta}{C_p T} \dot{Q} \qquad\qquad \dots (5.7)$$

$$\frac{\partial \theta}{\partial t} + u \frac{\partial \theta}{\partial x} + v \frac{\partial \theta}{\partial y} + \omega \frac{\partial \theta}{\partial p} = \frac{\theta}{C_p T} \dot{Q}$$

$$\theta = T \left(\frac{p_0}{p} \right)^{\kappa}$$

$$= \left(\frac{p_0}{p} \right)^{\kappa} \frac{p\alpha}{R}$$

$$= \left(\frac{p_0}{p} \right)^{\kappa} \frac{p}{R} \left(-\frac{\partial \phi}{\partial p} \right)$$

Substituting in equation (5.7)

$$\left(\frac{p_0}{p} \right)^{\kappa} \frac{p}{R} \left[\frac{\partial}{\partial t} \left(-\frac{\partial \phi}{\partial p} \right) + \vec{V} \cdot \nabla \left(-\frac{\partial \phi}{\partial p} \right) \right] + \omega \frac{\partial \theta}{\partial p} = \frac{\theta}{C_p T} \dot{Q}$$

Dividing by $\left(\dfrac{p_0}{p} \right)^{\kappa} \dfrac{p}{R}$

$$\frac{\partial}{\partial t} \left(-\frac{\partial \phi}{\partial p} \right) + \vec{V} \cdot \nabla \left(-\frac{\partial \phi}{\partial p} \right) + \frac{1}{\underset{\displaystyle T\alpha}{\theta T}} \omega \frac{\partial \theta}{\partial p} = \frac{\theta}{C_p T} \frac{1}{\underset{\displaystyle TR}{\theta p}} \dot{Q}$$

$$\frac{\partial}{\partial t} \left(-\frac{\partial \phi}{\partial p} \right) + \vec{V} \cdot \nabla \left(-\frac{\partial \phi}{\partial p} \right) + \frac{\alpha}{\theta} \frac{\partial \theta}{\partial p} \omega = \frac{R}{C_p p} \dot{Q}$$

$$\frac{\partial}{\partial t} \left(\frac{\partial \phi}{\partial p} \right) + \vec{V} \cdot \nabla \left(\frac{\partial \phi}{\partial p} \right) + \sigma \omega = -\frac{R}{C_p p} \dot{Q}$$

where $\sigma = -\dfrac{\alpha}{\theta} \dfrac{\partial \theta}{\partial p}$ is the static stability parameter.

The vertical scheme of levels is shown in figure 5.1.

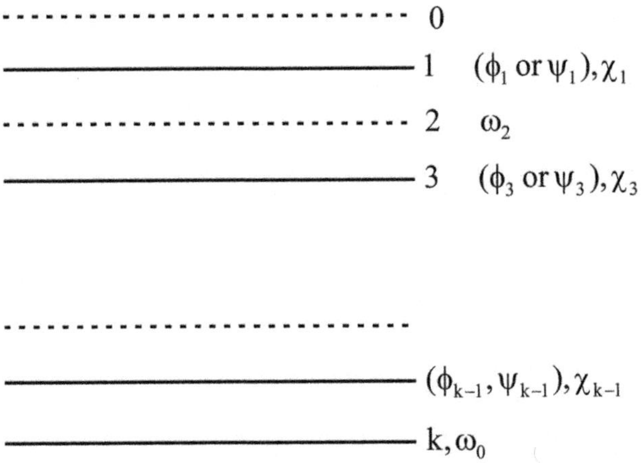

Fig. 5.1 Vertical scheme of levels

Quasi-geostrophic model (further discussions)

The vorticity equation in the quasi-geostrophic system is

$$\frac{\partial}{\partial t}\left(\frac{1}{f_0}\nabla^2\phi\right) + \vec{V}_g \cdot \nabla\left(\zeta_g + f\right) = f_0\frac{\partial\omega}{\partial p} \qquad \dots (5.8)$$

The thermodynamic energy equation is

$$\frac{\partial}{\partial t}\left(\frac{\partial\phi}{\partial p}\right) + \vec{V}_g \cdot \nabla\left(\frac{\partial\phi}{\partial p}\right) + \sigma\omega = -\frac{R}{C_p p}\dot{Q} \qquad \dots (5.9)$$

From this, we have

$$\omega = -\frac{1}{\sigma}\left\{\frac{\partial}{\partial t}\left(\frac{\partial\phi}{\partial p}\right) + \vec{V}_g \cdot \nabla\left(\frac{\partial\phi}{\partial p}\right) + \frac{R}{C_p p}\dot{Q}\right\}$$

Substituting in (5.8), we obtain

$$\frac{\partial}{\partial t}\left(\frac{1}{f_0}\nabla^2\phi\right) + \vec{V}_g \cdot \nabla\left(\zeta_g + f\right)$$

$$= -f_0\frac{\partial}{\partial p}\left[\frac{1}{\sigma}\left\{\frac{\partial}{\partial t}\left(\frac{\partial\phi}{\partial t} + \vec{V}_g \cdot \nabla\left(\frac{\partial\phi}{\partial p}\right) + \frac{R}{C_p p}\dot{Q}\right\}\right]$$

i.e.,
$$\frac{\partial}{\partial t}\left(\frac{1}{f_0}\nabla^2\phi + f_0\frac{\partial}{\partial p}\left(\frac{1}{\sigma}\frac{\partial\phi}{\partial p}\right)\right)+$$

$$+\vec{V_g}\cdot\nabla\left(\zeta_g + f + f_0\frac{\partial}{\partial p}\left(\frac{1}{\sigma}\frac{\partial\phi}{\partial p}\right)\right) = -\frac{f_0}{\sigma}\left(\frac{\partial}{\partial p}\frac{R}{C_p p}\dot{Q}\right)$$

In the adiabatic case we have $\dot{Q} = 0$

Therefore, $\dfrac{D\zeta_P}{Dt} = 0$

where $$\zeta_P = \zeta + f + \frac{\partial}{\partial p}\left(\frac{f_0}{\sigma}\frac{\partial\phi}{\partial p}\right) = \zeta_a + \frac{\partial}{\partial p}\left(\frac{f_0}{\sigma}\frac{\partial\phi}{\partial p}\right)$$

$$= \frac{1}{f_0}\nabla^2\phi + f + \frac{\partial}{\partial p}\left(\frac{f_0}{\sigma}\frac{\partial\phi}{\partial p}\right)$$

Thus ζ_P, the quasi-geostrophic potential vorticity is conserved in adiabatic frictionless large scale motion. Potential vorticity is absolute vorticity plus a thermal term. In the general case potential vorticity is generated by heating gradient and dissipated by friction.

Knowing the present geopotential field, the quasi-geostrophic potential vorticity equation can be used to predict the future ϕ field. This model is more general than the non-divergent barotropic model.

If we eliminate $\dfrac{\partial\phi}{\partial t}$ between equations (5.8) and (5.9) we obtain the quasi-geostrophic omega equation. This is a diagnostic equation in ω. Knowing the geopotential field and by solving the equation we can obtain the ω field.

The forcings on the right hand side are the vertical gradient of horizontal vorticity advection and the laplacian of horizontal thermal advection.

$$\left(\nabla^2 + \frac{f_0^2}{\sigma}\frac{\partial^2}{\partial p^2}\right)\omega = \frac{f_0}{\sigma}\frac{\partial}{\partial p}\left\{\vec{V_g}\cdot\nabla\left(\frac{1}{f_0}\nabla^2\phi + f\right)\right\}+$$

$$\frac{1}{\sigma}\nabla^2\left\{\vec{V_g}\cdot\nabla\left(-\frac{\partial\phi}{\partial p}\right)\right\}$$

The quasi-geostrophic model is widely used for theoretical studies.

5.5 Other Filtered Models

In the Linear Balance Model the divergence equation is

$$\nabla^2\phi - f\zeta + \beta u_\psi = 0$$

We retain terms of similar order in the vorticity equation, that is,

$$\frac{\partial\zeta}{\partial t} + \vec{V}_\psi \cdot \nabla(\zeta + f) + \vec{V}_\chi \cdot \nabla f = f\frac{\partial\omega}{\partial p}$$

where $\zeta = \nabla^2\psi$ and the thermodynamic energy equation is

$$\frac{\partial}{\partial t}\left(\frac{\partial\phi}{\partial p}\right) + \vec{V}\cdot\nabla\left(\frac{\partial\phi}{\partial p}\right) + \sigma\omega = -\frac{R}{C_p p}\dot{Q}$$

The continuity equation is

$$\nabla^2\chi + \frac{\partial\omega}{\partial p} = 0$$

In the Nonlinear balance model the divergence equation is

$$\nabla^2\phi - f\zeta + \beta u_\psi + 2J\left(v_\psi, u_\psi\right) = 0$$

The vorticity equation is

$$\frac{\partial\zeta}{\partial t} + \vec{V}\cdot\nabla(\zeta + f) + \omega\frac{\partial\zeta}{\partial p} = (\zeta + f)\frac{\partial\omega}{\partial p} - \hat{k}\cdot\nabla\omega\times\frac{\partial V_\psi}{\partial p}$$

and the thermodynamic energy equation is

$$\frac{\partial}{\partial t}\left(\frac{\partial\phi}{\partial p}\right) + \vec{V}\cdot\nabla\left(\frac{\partial\phi}{\partial p}\right) + \sigma\omega = -\frac{R}{C_p p}\dot{Q}$$

The continuity equation is of course

$$\nabla^2\chi + \frac{\partial\omega}{\partial p} = 0$$

These models are also referred to as filtered models. Because of the hydrostatic approximation sound waves are suppressed and because of the balance relation between the wind and pressure fields, gravity waves are filtered. Only the meteorologically significant Rossby waves are retained.

5.6   Numerical Methods

The equations we have come across are nonlinear partial differential equations. These are not amenable to analytical solution, in general. So we try to solve them numerically. The numerical methods employed are

  (i)   Finite difference method
  (ii)   Spectral method
  (iii)   Finite element method

We illustrate with the barotropic vorticity equation

$$\frac{\partial \zeta}{\partial t} + \vec{V}_\psi \cdot \nabla(\zeta + f) = 0$$

Adding $(\zeta + f)\nabla \cdot \vec{V}_\psi = 0$

We get,

$$\frac{\partial \zeta}{\partial t} + \nabla \cdot \vec{V}_\psi (\zeta + f) = 0$$

or $$\frac{\partial \zeta}{\partial t} = -\left\{ \frac{\partial}{\partial x}(u_\psi \zeta) + \frac{\partial}{\partial y}(v_\psi \zeta) + \beta v_\psi \right\}$$

We use a mesh of grid points $\delta x = \delta y = d$

$$x = m\delta x \ \ m = 0, 1, 2, 3, \ldots\ldots M$$

$$y = n\delta y \ \ n = 0, 1, 2, 3, \ldots\ldots N$$

$$u_\psi = u_{m,n} = -\frac{\Psi_{m,\,n+1} - \Psi_{m,\,n-1}}{2d}$$

$$v_\psi = v_{m,n} = \frac{\Psi_{m+1,\,n} - \Psi_{m-1,\,n}}{2d}$$

in centered difference.

Similarly,

$$\zeta_{m,n} = \nabla^2 \psi = \frac{\partial^2 \psi}{\partial x^2} + \frac{\partial^2 \psi}{\partial y^2}$$

$$= \frac{\psi_{m+1,n} + \psi_{m-1,n} + \psi_{m,n+1} + \psi_{m,n-1} - 4\psi_{m,n}}{d^2}$$

The non-divergent barotropic vorticity equation in finite difference form is

$$\frac{\zeta_{m,n}(t+\delta t) - \zeta_{m,n}(t-\delta t)}{2\delta t} =$$

$$\frac{1}{2d}\left[\frac{\left(u_{m+1,n}\,\zeta_{m+1,n} - u_{m-1,n}\,\zeta_{m-1,n}\right)}{+\left(v_{m,n+1}\,\zeta_{m,n+1} - v_{m,n-1}\,\zeta_{m,n-1}\right)}\right] + \beta\, v_{m,n}$$

We have one such equation for each grid point. Thus we have a set of simultaneous difference equations.

In the spectral method the barotropic vorticity equation is written in spherical polar coordinates.

$$\frac{\partial}{\partial t}\nabla^2 \psi = \frac{1}{a^2}\left[\frac{\partial \psi}{\partial \mu}\frac{\partial}{\partial \lambda}\nabla^2 \psi - \frac{\partial \psi}{\partial \lambda}\frac{\partial}{\partial \mu}\nabla^2 \psi\right] - 2\frac{\Omega}{a^2}\frac{\partial \psi}{\partial \lambda}$$

Where λ is the longitude and $\mu = \sin\phi$ where ϕ is the latitude. The stream function is expanded in terms of spherical harmonics

$$\psi(\lambda, \mu, t) = a^2 \sum_m \sum_n \psi_{m,n}(t)\, Y_{m,n}(\mu, \lambda)$$

$$= \sum \sum \psi_{m,n}(t)\, P_{m,n}\, e^{im\lambda}$$

We obtain one equation for each $\psi_{m,n}$. Thus we get a set of simultaneous equations.

5.7   Subgrid Scale Processes

Friction

$$\frac{d\overline{u}}{dt} = -\alpha\frac{\partial \overline{p}}{\partial x} + f\,\overline{v} + \frac{\partial}{\partial z}\left(-\overline{u'w'}\right)$$

The meteorological observations like wind represent an average for a few minutes (bar on top).

$$u = \bar{u} + u'$$

In the planetary boundary layer (about 1 km from the lower boundary i.e., land or ocean) turbulent motion is predominant. Turbulent eddies of various scales transport momentum, heat, moisture or pollutants in the vertical direction. The simplest formulations for this vertical transfer are the so called bulk formulae.

For momentum flux

$$\tau_s = - \rho C_D V_s V_s$$

where C_D is the drag coefficient and V_s is the wind speed.

Heat transfer or flux $H_s = \rho C_p C_H V_s (T_s - T_o)$

T_s is the surface temperature, T_o is air temperature at the ship deck level.

Evaporation $E_w = \rho C_E V_s (q_s - q_o)$.

C_H and C_E are coefficients similar to C_D, q_s is the saturation mixing ratio and q_o is mixing ratio at ship deck level.

In the thermodynamic energy equation, the diabatic heating.

$$\dot{Q} = \dot{Q}_R + \dot{Q}_s + \dot{Q}_C$$

where $\dot{Q}_R$ - radiative heating

            $\dot{Q}_s$ - sensible heating

            $\dot{Q}_C$ - Convective heating

The radiative heating

$$\dot{Q}_R = \dot{Q}_{solar} + \dot{Q}_{terrestrail}$$

The sun radiates at a surface temperature of 6000 °K. This solar or short wave radiation ranges from 0.1 μ to 5 μ and peaks in green. This consists of visible, ultraviolet and infrared radiation. The atmosphere is almost transparent to this radiation. Only ultraviolet radiation is absorbed by ozone in the stratosphere. There is however molecular and mie scattering. The solar radiation is reflected from the earth – more by clouds, snow, sand. This reflected radiation is known as albedo. The earth-atmosphere system radiates at around 250 °K in the near and far infrared. This terrestrial or long wave radiation ranges for 2 μ to 100 μ.

The earth's atmosphere is rather opaque to terrestrial radiation. Water vapour, carbon dioxide, methane etc., absorb and re-radiate terrestrial radiation. Atmospheric models calculate seasonal and diurnal as well as latitudinal variation of solar radiation, its scattering and reflection. They also calculate the absorption and re-radiation of terrestial radiation.

Cumulus heating

The tropical atmosphere is conditionally unstable i.e., it is stable so long as it is dry or unsaturated and unstable when it is saturated. The resulting moist convection is of scale 10 km. There are formulations to compute the latent heat released in a population of cumulus clouds. This is known as parameterization of cumulus heating. This cumulus heating is important for tropical depressions, storms and monsoons.

5.8   Simulation of the Atmospheric General Circulation

We illustrate the power of numerical models with the results of one of the very sophisticated models - IITM earth system model (ESM).These were kindly provided by Dr. A.D. Choudhury of CCCS, IITM, Pune (P. Swapna, R. Krishnan, N. Sandeep, A.G. Prajeesh, D.C. Ayantika, S. Manmeet and R. Vellore, 2018).

These models are the result of years of work of several scientists. First there were atmospheric general circulation models. Then came atmosphere-ocean coupled models .Now we have the much more comprehensive earth system models. Starting from the input of solar radiation the atmospheric general circulation is simulated .This should be considered, in my opinion, one of the great scientific achievements of atmospheric scientists.

The IITM ESM model has atmosphere, land, sea-ice, ocean and interactive bio-geochemistry. The atmospheric part is a spectral general circulation model.

The maps shown are results of long-term (century) present day climate integrations. In addition to simulating mean climate, these models are used to study climate variability as also impact of increased green house gases.

Fig. 5.2b shows the 850 hpa simulated wind field during July (northern summer). The realistic monsoon southwesterly/ westerly winds over India and southeast Asia are very well reproduced. The turning of the southeast trades into monsoon winds can also be seen. The north Pacific and north Atlantic anticyclones can be noticed.

Fig. 5.2a shows the simulated 850 hpa wind field for January (northern winter). The northeast trade winds over equatorial north Pacific and north Atlantic can be seen. The Siberian high over Asia, the lows over north Pacific and north Atlantic can be seen.

Fig 5.3a shows the simulated upper tropospheric wind field at 200 hpa during northern winter (January).The strong midlatitude westerly jet stream notably in the northern (winter) hemisphere is very well simulated.

Next we look at the upper tropospheric wind field during northern summer (July) Fig. 5.3b. The notable features are the easterly jet stream over India and neighbourhood and the Tibetan anticyclone. The weakened westerly jet stream over northern hemisphere (also shifted northwards) and the strengthened southern hemisphere jet stream can be seen. These features are all realistically simulated.

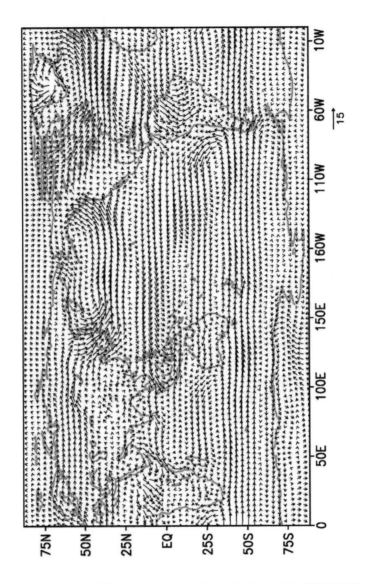

Fig. 5.2a Jan 850 hPa present day climatological winds (IITM-ESM)

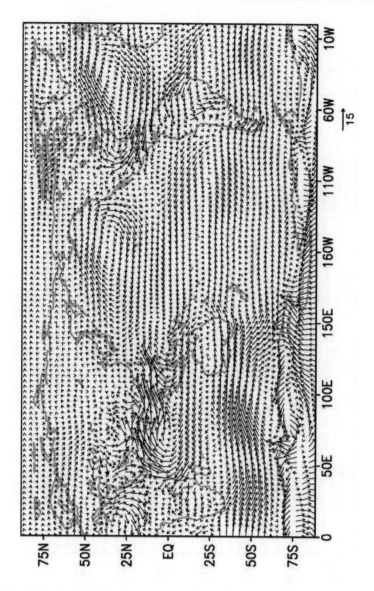

Fig. 5.2b Jul 850 hPa present day climatological winds (IITM-ESM)

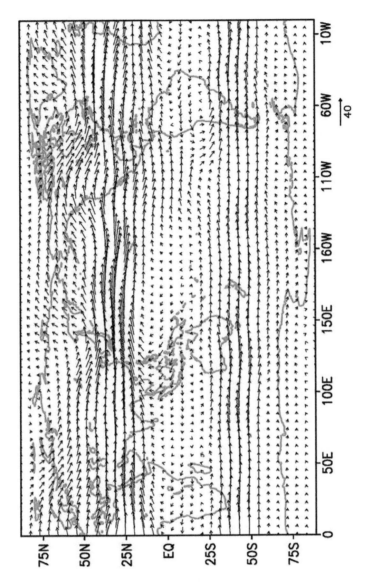

Fig. 5.3a Jan 200 hPa present day climatological winds (IITM-ESM)

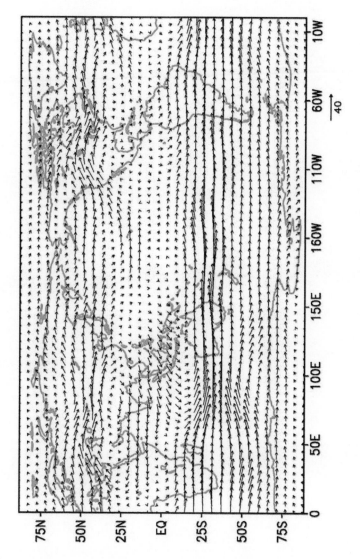

Fig. 5.3b July 200 hPa present day climatological winds (IITM-ESM)

6. Hydrodynamic Instability

6.1 Introduction

Generally atmospheric flows exhibit horizontal or vertical shear or both. In general the zonal flow U is a function of y or p or both i.e., $U(y, p)$.

Consider a basic flow U and a wave perturbation $u' = u\exp(ik(x-ct))$ superimposed on it.

c is the phase speed (eastward positive) of the wave.

If the wave perturbation grows spontaneously drawing energy (feeding on) from the basic flow, the basic flow U is said to be (hydrodynamically) unstable. A classic example of this is Rayleigh instability.

Usually atmospheric flows with shear i.e., meridional or vertical or both i.e., $U(y, p)$, exhibit hydrodynamic instability under certain conditions. Clearly this is very different from hydrostatic stability or instability of a parcel of air.

Middle latitude westerly flows, tropical easterly flows and monsoon flows have meridional or vertical shear or both.

6.2 Barotropic Instability

If the basic flow has only horizontal (meridional) shear i.e., $\dfrac{\partial U}{\partial y}$ and if a superimposed wave perturbation grows spontaneously, then the basic flow is said to be barotropically unstable. Here the perturbation feeds on the kinetic energy of the basic flow and grows.

Consider the non-divergent barotropic vorticity equation

$$\frac{d\zeta_a}{dt} = 0$$

$$\frac{\partial \zeta_a}{\partial t} + \vec{V_g} \cdot \nabla(\zeta + f) = 0 \qquad \qquad .....(6.1)$$

$$\frac{\partial \zeta}{\partial t} + u \frac{\partial \zeta}{\partial x} + v \frac{\partial \zeta}{\partial y} + \beta v = 0 \qquad .....(6.2)$$

Basic flow $U(y)$

$$U = U + u', \quad \overline{v} = 0, \quad v = 0 + v'$$

The perturbation barotropic vorticity equation is

$$\left(\frac{\partial}{\partial t} + U \frac{\partial}{\partial x}\right) \zeta' + u' \frac{\partial \zeta'}{\partial x} + v' \frac{\partial \zeta'}{\partial y} + v' \frac{\partial \overline{\zeta}}{\partial y} + \beta v' = 0 \qquad .....(6.3)$$

Neglecting second order terms, we obtain

$$\left(\frac{\partial}{\partial t} + U \frac{\partial}{\partial x}\right) \zeta' + v' \left(\beta - \frac{\partial^2 U}{\partial y^2}\right) = 0$$

$$\left(\frac{\partial}{\partial t} + U \frac{\partial}{\partial x}\right) \nabla^2 \psi' + \frac{\partial \psi'}{\partial x} \left(\beta - \frac{\partial^2 U}{\partial y^2}\right) = 0 \qquad .....(6.4)$$

Assume wave perturbations of the type

$$\psi' = \psi e^{ik(x-ct)}$$

Substituting, we obtain

$$ik(U - c)\left(-k^2 \psi + \frac{\partial^2 \psi}{\partial y^2}\right) + ik\psi\left(\beta - \frac{\partial^2 U}{\partial y^2}\right) = 0$$

$$(U - c)\left(\frac{\partial^2 \psi}{\partial y^2} - k^2 \psi\right) + \left(\beta - \frac{\partial^2 U}{\partial y^2}\right)\psi = 0$$

It was shown (Kuo, 1949) that for the perturbation to grow i.e., $c = c_r + c_i$ and $c_i \neq 0$,

the necessary condition is

$$\int_0^D |\psi|^2 \frac{\beta - \dfrac{\partial^2 U}{\partial y^2}}{|U - c|^2} \, dy = 0$$

This can happen only if $\beta - \dfrac{\partial^2 U}{\partial y^2}$ i.e., $\dfrac{\partial \overline{\zeta}_a}{\partial y}$

i.e., the meridional gradient of basic absolute vorticity changes sign somewhere in the channel 0 to D.

This is a generalization of the Rayleigh Instability condition where there was no rotation.

6.3   Baroclinic Instability

If the basic flow has vertical shear $\dfrac{\partial U}{\partial p}$ and if a (small) wave perturbation is superimposed on it, it has been shown (Charney, 1947, Eady, 1949) that the perturbation can grow at the expense of the basic flow under certain conditions. Then the basic flow is said to be baroclinically unstable and the disturbance is a baroclinically growing disturbance. Middle latitude westerly flow exhibits baroclinic instability. Here the wave perturbation feeds on the available potential energy of the basic flow arising from the south to north temperature contrast. The zonal available potential energy is converted into eddy available potential energy and then into eddy kinetic energy.

This can be explained with a two-level quasi-geostophic model. (Fig. 6.1).

$$\frac{\partial}{\partial t}\left(\nabla^2 \psi_1\right) + \vec{V}_1 \cdot \nabla\left(\nabla^2 \psi_1\right) + \beta \frac{\partial \psi_1}{\partial x} = \frac{f_0}{\Delta p}\omega_2 \qquad \dots (6.5)$$

$$\frac{\partial}{\partial t}\left(\nabla^2 \psi_3\right) + \vec{V}_3 \cdot \nabla\left(\nabla^2 \psi_3\right) + \beta \frac{\partial \psi_3}{\partial x} = -\frac{f_0}{\Delta p}\omega_2 \qquad \dots (6.6)$$

$$\frac{\partial}{\partial t}\left(\psi_1 - \psi_3\right) + \vec{V}_2 . \nabla\left(\psi_1 - \psi_3\right) = \frac{\sigma \Delta p}{f_0}\omega_2 \qquad \dots (6.7)$$

Linearize the above equations.

The perturbation vorticity equations are written at levels 1 and 3 and the perturbation thermodynamic energy equation is written at level 2 (Fig. 6.1). Eliminating ω_2' we obtain perturbation potential vorticity equations.

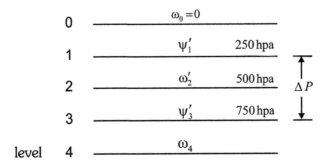

Fig. 6.1 Vertical level scheme in a two level model

Assuming wave solutions and substituting, we get two algebraic equations in two unknowns ψ_1' and ψ_3'. For nontrivial solutions the determinant of the coefficients should vanish. This yields a quadratic equation in c. If the discriminant is negative then c is complex and there is a possibility of growth.

Charney and Stern (1962) showed that in a flow with meridional and vertical shear i.e., $U(y, p)$ there is a possibility of the internal jet being unstable if the meridional gradient of basic potential vorticity $\left(\dfrac{\partial \bar{\zeta}_P}{\partial y} \right)$ goes through zero somewhere in the channel. Keshavamurty et.al. (1978) showed that this condition is satisfied in the monsoon region. Satyan et.al. (1980) conducted combined barotropic-baroclinic instability analysis of monsoon flow using a two level quasi-geotrophic model and showed that monsoon disturbances can grow by this process. Goswami et.al. (1980) showed that lower tropospheric monsoon flow is barotropically unstable.

6.4 Sources of Energy for Atmospheric Disturbances

Middle latitude disturbances like extra-tropical depressions feed on horizontal temperature gradient or contrast between warm and cold air masses. Thermal advection is important. Middle latitude westerly flow exhibits baroclinic instability. These disturbances derive their energy from zonal available potential energy arising out of south to north temperature contrast and this is converted to eddy available potential energy which is converted to eddy kinetic energy. Here $\dfrac{\partial U}{\partial p}$ or $\dfrac{\partial T}{\partial y}$ is important.

Fig. 6.2 (a, b, c, d) show wind flow pattern and geopotential field of a winter disturbance (31 Jan 2019) over northwest India, Pakistan and neighbourhood. These are known as western disturbances as they move from the west. They are structurally and energetically similar to midlatitude disturbances. The associated cyclonic circulation (in this case) extends up to 500 hpa. They cause rain/snowfall over western Himalayas and neighbourhood. Notice also the associated (Fig. 6.2e) westerly trough intruding into low latitudes over India. The clouds associated with a western disturbance are seen in Fig. 6.3.

Tropical lower tropospheric flows, on the other hand, appear to be barotropically unstable. Here the disturbances feed on the kinetic energy of the basic flow. Here the meridional shear $\dfrac{\partial U}{\partial y}$ is important. Horizontal temperature gradients are weaker and thermal advection is less pronounced.

Monsoon flows exhibit both horizontal (meridional) and vertical shear. U is a function of both y and z (or p). Thus scientists have studied combined barotropic–baroclinic instability of monsoon flows.

There is another important mechanism for growth of tropical disturbances. Tropical soundings lie between dry adiabatic lapse rate (DALR) and saturation adiabatic lapse rate (SALR). Thus tropical atmosphere is conditionally unstable i.e., it is stable for ascent of dry or unsaturated air but unstable for ascent of saturated air. Theoretical studies show that in such an atmosphere the unstable mode that grows fastest is the ten kilometer cumulus scale. However, observations show that hundred to thousand kilometer scale disturbances like tropical cyclones and monsoon depressions grow. If these disturbances compete with the cumulus scale, the cumulus scale will win and grow. Thus, Charney and Eliassen (1964) and others proposed that the synoptic scale and the cumulus scale co-operate with each other rather than compete. The large scale provides low level frictional convergence and import of moisture for the cumulus scale to grow. The population or ensemble of cumulus clouds releases latent heat of condensation in the region of the low pressure area in contrast to the surroundings and thus provides the differential heating to drive and strengthen the depression (heat) engine. They called this conditional instability of the second kind or CISK. Tropical cyclones, hurricanes, typhoons grow by this process.

Fig. 6.3 shows INSAT (infrared) cloud pictures on 7, Feb 2019. Notice the two tropical cyclones in the tropical Indian ocean east of Madagascar (southern summer). These are, of course, not as impressive or menacing as northwest Pacific typhoons. Typhoons of the pacific, hurricanes of the Atlantic and cyclones of the Indian Ocean are (cousins) the same species.

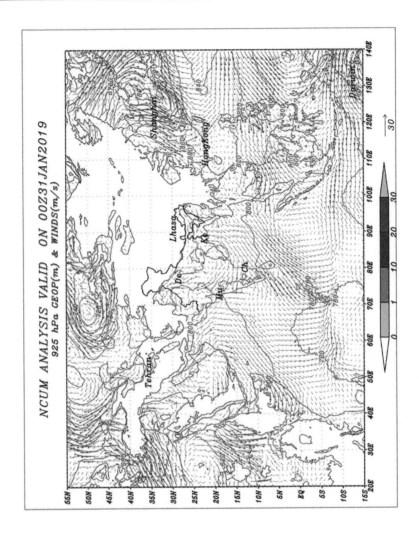

Fig. 6.2a Wind flow at 925 hpa, 31st Jan 2019

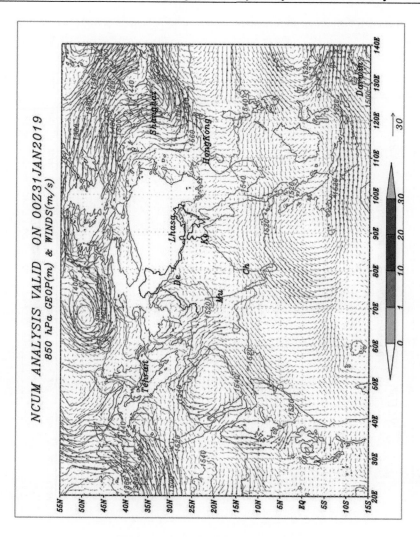

Fig. 6.2b Wind flow at 850 hpa, 31st Jan 2019

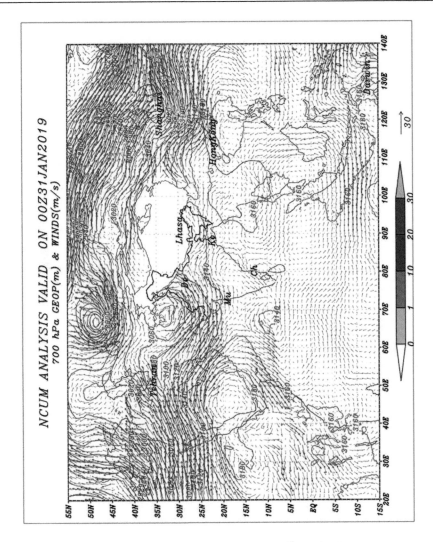

Fig. 6.2c Wind flow at 700 hpa, 31st Jan 2019

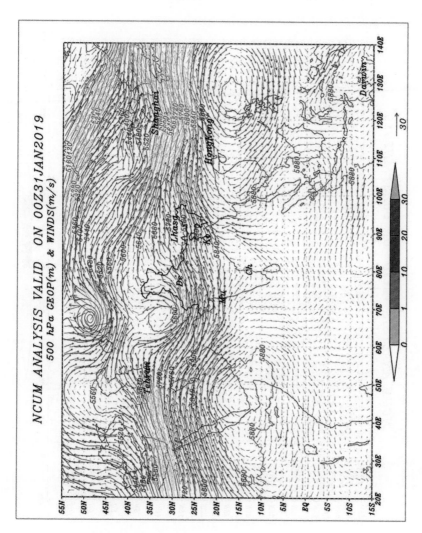

Fig. 6.2d Wind flow at 500 hpa, 31st Jan 2019

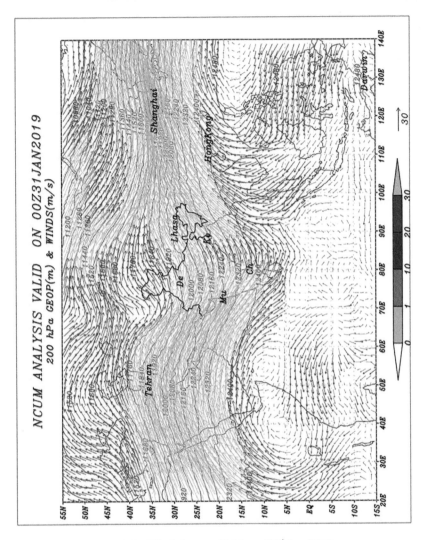

Fig. 6.2e Wind flow at 200 hpa, 31st Jan 2019

SAT :INSAT–3D IMG 07-02-2019/(1530 to 1556) GMT
IMG_TIR1 10.8 um 07-02-2019/(2100 to 2126) IST
L1B FULL DISK (LINEAR STRETCH: 1.0%)

IMD/Delhi
527 898

Fig. 6.3 INSAT Cloud Picture 7 Feb 2019

Figures 6.4 a, b and 6.4c show wind flow patterns at 925 hpa, 850 hpa and 700 hpa respectively. Notice the tropical depression over tropical northwest Pacific near Philippines and extra-tropical disturbances over the northwest and northeast corners of the map.

Monsoon depressions appear to be a hybrid between mid-latitude depressions and tropical cyclones. Here the shears of the basic flow and release of latent heat in cumulus convection both appear to be important. Fig. 6.5a to Fig. 6.5j show the wind field in and around a typical monsoon depression. The cyclonic circulation associated with the monsoon depression extends up to 400 hpa with a southwestward tilt with height. These form over head Bay of Bengal and move westnorthwestwards across central and adjoining north India. They are associated with heavy rainfall in the southwest sector. Monsoon lows and depressions are the main rain producing disturbances during the (summer) southwest monsoon season.

There is another important class of monsoon disturbances called the mid-tropospheric cyclone (MTC).This has maximum intensity in the middle troposphere i.e 700 to 500 hpa. This was discovered (Miller F.R and Keshavamurthy R.N, 1968) during the meteorology program of the International Indian Ocean Expedition. This occurs over northeast Arabian sea and adjoining areas. During many active monsoon epochs a monsoon low or depression over north Bay of Bengal and an MTC over northeast Arabian sea occur simultaneously and are embedded in east-west shear zone in the middle troposphere. Choudhury, A.D et.al., (2018) studied the dynamics of these disturbances and stressed the role of sheet type clouds.

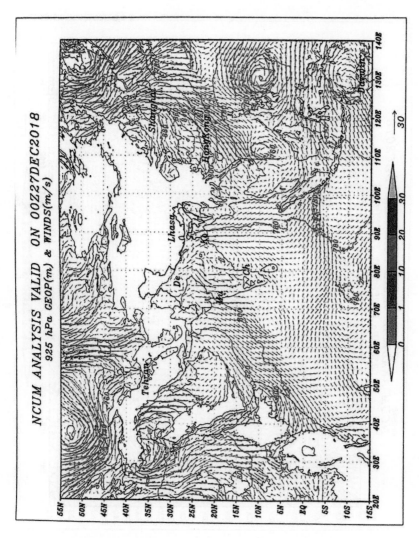

Fig. 6.4a Wind flow at 925 hpa, 27th December, 2018

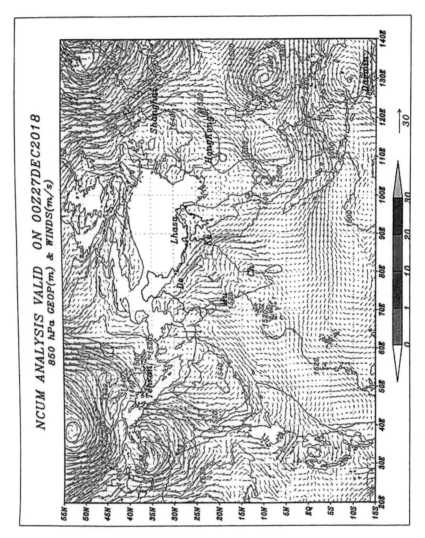

Fig. 6.4b Wind flow at 850 hpa, 27th December, 2018

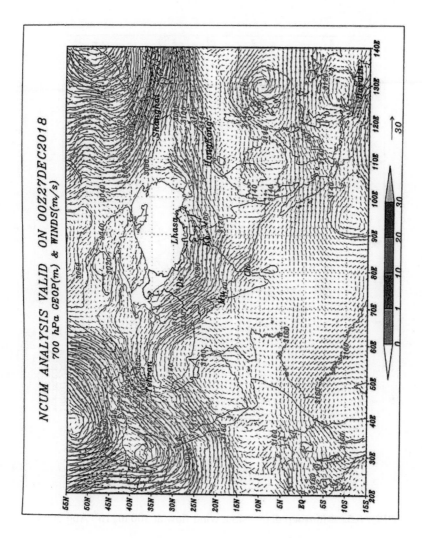

Fig. 6.4c Wind flow at 700 hpa, 27th December, 2018

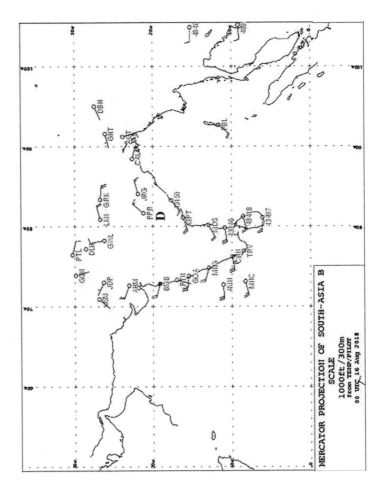

Fig. 6.5a Monsoon depression, wind flow at 300 m, 16th Aug 2018

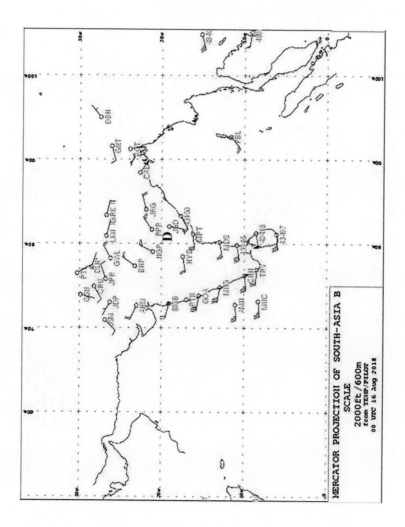

Fig. 6.5b Monsoon depression, wind flow at 600 m, 16th Aug 2018

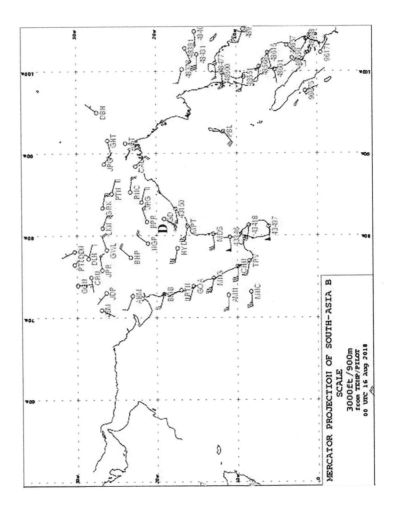

Fig. 6.5c Monsoon depression, wind flow at 900 m, 16th Aug 2018

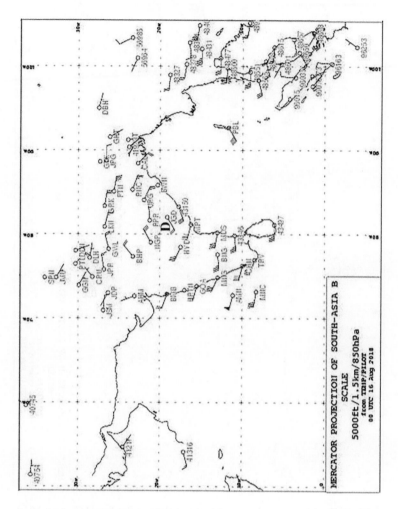

Fig. 6.5d Monsoon depression, wind flow at 850 hPa, 16th Aug 2018

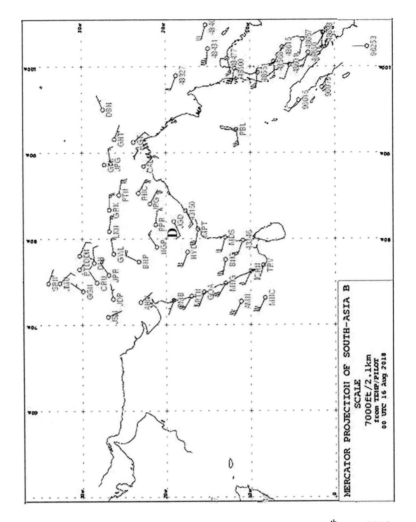

Fig. 6.5e Monsoon depression, wind flow at 2.1 km, 16th Aug 2018

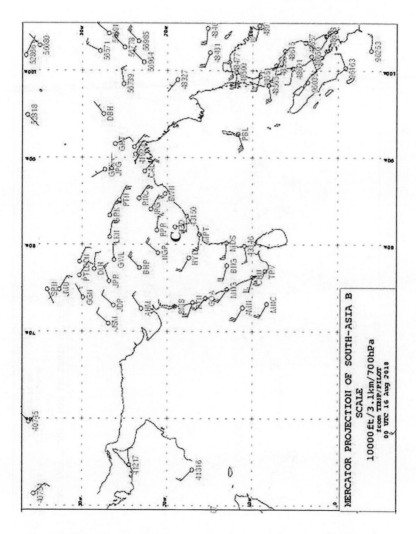

Fig. 6.5f Monsoon depression, wind flow at 700 hpa, 16th Aug 2018

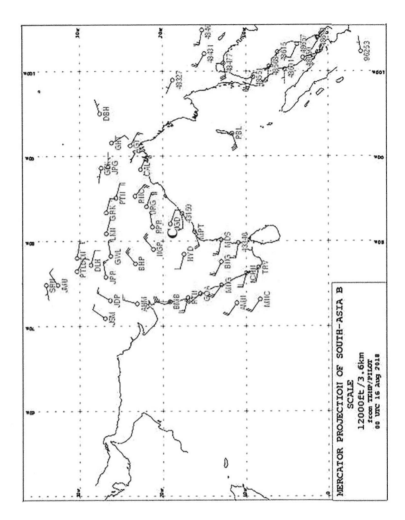

Fig. 6.5g Monsoon depression, wind flow at 3.6 km, 16th Aug 2018

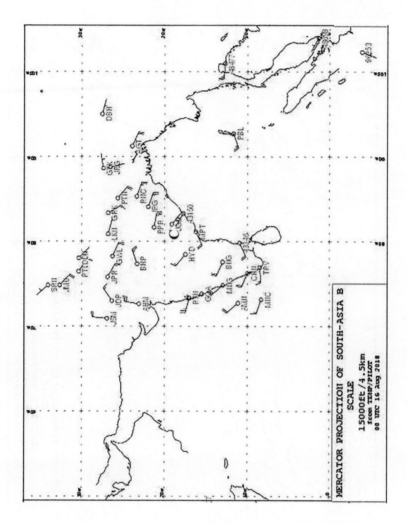

Fig. 6.5h Monsoon depression, wind flow at 4.5 km, 16th Aug 2018

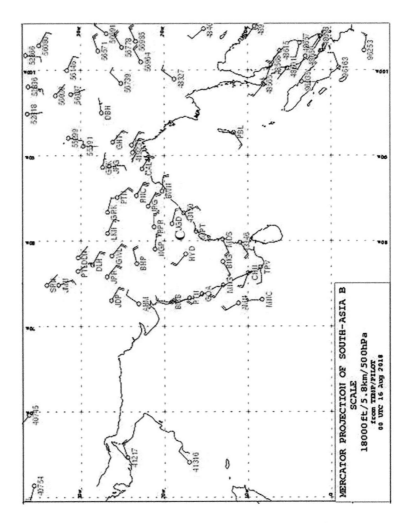

Fig. 6.5i Monsoon depression, wind flow at 500 hpa, 16th Aug 2018

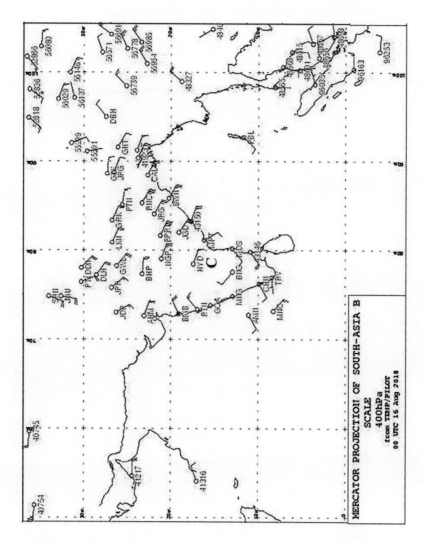

Fig. 6.5j Monsoon depression, wind flow at 400 hpa, 16th Aug 2018

Appendix 1

Pressure Gradient Force

Consider a fixed volume $\delta x\, \delta y\, \delta z$ (Fig. A1.1)

Pressure gradient force

Force on face A

$$= \left(p - \frac{\partial p}{\partial x} \frac{\delta x}{2} \right) \delta y\, \delta z$$

Force on face B

$$= -\left(p + \frac{\partial p}{\partial x} \frac{\delta x}{2} \right) \delta y\, \delta z$$

Net pressure force in x direction $= -\dfrac{\partial p}{\partial x} \delta x\, \delta y\, \delta z$

Dividing by the mass $\rho\, \delta x\, \delta y\, \delta z$, we get the net pressure force per unit mass in the x direction as $-\dfrac{1}{\rho} \dfrac{\partial p}{\partial x} = -\alpha \dfrac{\partial p}{\partial x}$.

Considering all the three directions, the net pressure force per unit mass is

$$= -\alpha \left(\frac{\partial p}{\partial x} \hat{i} + \frac{\partial p}{\partial y} \hat{j} + \frac{\partial p}{\partial z} \hat{k} \right) = -\alpha\, \nabla p$$

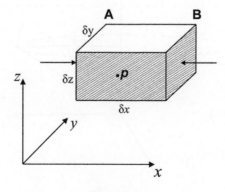

Fig. A1.1

Appendix 2

Equation of Motion in Rotating Co-ordinates

Newton's second law of motion is valid in an absolute frame of reference. We consider an absolute frame fixed at the centre of the earth. What we observe on the earth i.e., wind, cloud motion etc., is with reference to the rotating earth. So we consider another frame rotating with the earth.

$\vec{r}$ is a position vector

The velocity $\vec{V} = \dfrac{d\vec{r}}{dt}$

The acceleration $\dfrac{d\vec{V}}{dt} = \dfrac{d^2\vec{r}}{dt^2}$

$\vec{\Omega}$ is the angular velocity (of the earth)

Linear velocity $\vec{V} = \vec{\Omega} \times \vec{r}$

It is perpendicular to both $\vec{\Omega}$ and $\vec{r}$. (Fig. A2.1)

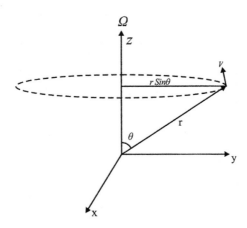

Fig. A2.1

Let us consider an absolute co-ordinate system (x, y, z). We have another co-ordinate system (x', y', z') with the same origin but rotating with respect to the fixed frame with an angular velocity $\vec{\Omega}$.

Any vector $\vec{F} = F_x \hat{i} + F_y \hat{j} + F_z \hat{k}$
$$= F_x' \hat{i}' + F_y' \hat{j}' + F_z' \hat{k}'$$

It is to be noted that, $\hat{i}'$, $\hat{j}'$ and $\hat{k}'$ are functions of time i.e., they are changing with time.

$$\frac{d\vec{F}}{dt} = \frac{dF_x}{dt}\hat{i} + \frac{dF_y}{dt}\hat{j} + \frac{dF_z}{dt}\hat{k}$$

$$= \frac{dF_x'}{dt}\hat{i}' + \frac{dF_y'}{dt}\hat{j}' + \frac{dF_z'}{dt}\hat{k}' + F_x'\frac{d\hat{i}'}{dt} + F_y'\frac{d\hat{j}'}{dt} + F_z'\frac{d\hat{k}'}{dt}$$

Now,

$$\frac{d\hat{i}'}{dt} = \vec{\Omega} \times \hat{i}'$$

$$\frac{d\hat{j}'}{dt} = \vec{\Omega} \times \hat{j}'$$

$$\frac{d\hat{k}'}{dt} = \vec{\Omega} \times \hat{k}'$$

Therefore $\dfrac{d\vec{F}}{dt}$ (in absolute frame)

$$= \frac{d\vec{F}}{dt} \text{ (in rotating frame)} + \vec{\Omega} \times \vec{F}'$$

Thus, $\dfrac{d_a\vec{r}}{dt} = \dfrac{d\vec{r}}{dt} + \vec{\Omega} \times \vec{r}$

or, $\vec{V}_a = \vec{V} + \vec{\Omega} \times \vec{r}$

Then, $\dfrac{d_a\vec{V}_a}{dt} = \dfrac{d\vec{V}_a}{dt} + \vec{\Omega} \times \vec{V}_a$

$$= \frac{d}{dt}\left(\vec{V} + \vec{\Omega} \times \vec{r}\right) + \vec{\Omega} \times \left(\vec{V} + \vec{\Omega} \times \vec{r}\right)$$

$$= \frac{d\vec{V}}{dt} + 2\vec{\Omega} \times \vec{V} - \Omega^2 \vec{r}$$

Thus the equation of motion in a co-ordinate system rotating with the earth is

$$\frac{d\vec{V}}{dt} = -\alpha \nabla p - 2\vec{\Omega} \times \vec{V} - \Omega^2 \vec{r} + \vec{g} + \vec{F}$$

$$= -\alpha \nabla p - 2\vec{\Omega} \times \vec{V} + \vec{g}_a + \vec{F}$$

where $\vec{g}_a$ is the gravitational acceleration plus the centrifugal acceleration.

$-2\vec{\Omega} \times \vec{V}$ is the Coriolis force due the rotation of the earth.

Consider a local co-ordinate system on a tangential plane. (Fig. A2.2).

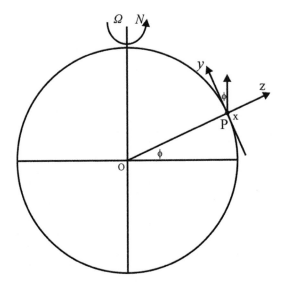

Fig. A2.2

We use a Cartesian coordinate system at any point P on the surface of the earth. We consider a plane tangential to the earth's surface at point P. On this plane x axis points to the east (into the plane of the paper), y axis points to the north and z axis upwards. This system is fixed to the rotating earth.

The equation of motion is

$$\frac{d\vec{V}}{dt} = -\alpha \, \nabla p - 2\vec{\Omega} \times \vec{V} + \vec{g} + \vec{F}$$

Let us consider the coriolis term $-2\vec{\Omega} \times \vec{V}$

$$\Omega_y = \Omega \cos \phi$$

$$\Omega_z = \Omega \sin \phi$$

$$\Omega_x = 0$$

$$-2\vec{\Omega} \times \vec{V} = -2\Omega \begin{bmatrix} \hat{i} & \hat{j} & \hat{k} \\ 0 & \cos\phi & \sin\phi \\ u & v & w \end{bmatrix}$$

Thus, the component equations are:

$$\frac{du}{dt} = -\alpha \frac{\partial p}{\partial x} + 2\Omega \sin \phi . v - 2\Omega \cos \phi \cdot w + F_x$$

$$\frac{dv}{dt} = -\alpha \frac{\partial p}{\partial y} - 2\Omega \sin \phi \cdot u + F_y$$

$$\frac{dw}{dt} = -\alpha \frac{\partial p}{\partial z} - g + 2u \cos \phi \cdot \Omega + F_z$$

Denoting $2\Omega \sin \phi = f$, the Coriolis parameter and neglecting the smaller order $\cos \phi$ terms, we obtain,

$$\frac{du}{dt} = -\alpha \frac{\partial p}{\partial x} + f v + F_x$$

$$\frac{dv}{dt} = -\alpha \frac{\partial p}{\partial y} - f u + F_y$$

$$\frac{dw}{dt} = -\alpha \frac{\partial p}{\partial z} - g + F_z$$

Appendix 3

Equation of State for Air (a mixture of gases)

The atmospheric air is a mixture of gases. We can now derive an expression for the equation of state for a mixture of gases. According to Dalton's law each constituent perfect gas occupies the whole volume completely. Each constituent (perfect) gas obeys equation of state. The total pressure exerted by the mixture of gases is the sum of the partial pressures of the individual gases. Thus, for example,

$$p_{N_2} V = M_{N_2} R_{N_2} T$$

where p_{N_2} is the partial pressure of Nitrogen gas. So the sum total for all constituent gases is

$$p_{N_2} V + p_{O_2} V + p_{Ar} V + p_{CO_2} V + \ldots\ldots = pV$$

The total mass $M = \Sigma M_i$

Dividing by M

$$p\frac{V}{M} = p\alpha = RT, \quad \alpha = \frac{1}{\rho}$$

R is the gas constant for the mixture of gases (i.e., dry air).

$$R = \frac{M_{N_2}}{M} R_{N_2} + \frac{M_{O_2}}{M} R_{O_2} + \frac{M_{Ar}}{M} R_{Ar} + \frac{M_{CO_2}}{M} R_{CO_2} + \ldots$$

Appendix 4

Rossby Number

Consider the horizontal equation of motion

$$\frac{\partial u}{\partial t} + u\frac{\partial u}{\partial x} + v\frac{\partial u}{\partial y} + w\frac{\partial u}{\partial z} = -\alpha\frac{\partial p}{\partial x} + fv - 2w\Omega\cos\phi + F_x$$

$$\frac{\partial u}{\partial t} \sim \frac{U}{t} \sim \frac{U}{\dfrac{L}{U}} \sim \frac{U^2}{L}$$

where U is the order of magnitude of horizontal wind

L is the length scale

t is the time scale

$$u\frac{\partial u}{\partial x} \sim \frac{U^2}{L}$$

$$fu \sim fv \sim fU$$

$$\frac{\left|\dfrac{\partial u}{\partial t}\right|}{|fu|} = \frac{\dfrac{U^2}{L}}{fU} = \frac{U}{fL} = R_0 , \text{ the Rossby number.}$$

Rossby number is the ratio of inertial forces to Coriolis force. This is a dimensionless number.

Mid-latitudes

Latitude 45° N

$$f \sim 10^{-4}\,\text{s}^{-1}$$

$$L \sim 10,000 \text{ km} \quad R_0 = 0.01$$

$$L \sim 1,000 \text{ km} \qquad R_0 = 0.1$$

$$L \sim 100 \text{ km} \qquad R_0 = 1$$

$$L \sim 10 \text{ km} \qquad R_0 = 10$$

Low Latitude

Latitude $10° \text{ N}$

$$f \sim 2.5 \times 10^{-5} \text{s}^{-1}$$

$$L \sim 10,000 \text{ km} \quad R_0 = 0.04$$

$$L \sim 1,000 \text{ km} \qquad R_0 = 0.4$$

$$L \sim 100 \text{ km} \qquad R_0 = 4$$

$$L \sim 10 \text{ km} \qquad R_0 = 40$$

For large scale and synoptic scale motions the Rossby number is small, especially in middle latitudes. For these small Rossby number regimes, the motion is quasi-geostrophic and rotation dominated.

For smaller scales – sea and land breeze, urban scale, meso-scale, orographic (small and medium) scale etc., the Rossby number is large and non-geostrophic effects are important.

Appendix 5

Dry Adiabatic Lapse Rate (DALR)

The first law of thermodynamics is

$$dQ = C_p dT - \alpha dp$$

For an adiabatic process, we have

$$0 = C_P \, dT - \alpha \, dp$$

Dividing by T

$$0 = C_P \frac{dT}{T} - R \frac{dp}{p} \qquad \text{since } p\alpha = RT$$

Substitute $\quad gdz = -\alpha dp \quad$ since $\quad dp = -g\rho dz$

Therefore, $\quad C_p dT + g \, dz = 0$

So $\qquad -\dfrac{\partial T}{\partial z} = \dfrac{g}{C_P} = \dfrac{g}{C_{pd}} = \gamma_d = 9.8\ ^\circ\text{C/km}$

Similarly for saturated air

$$\gamma_s \approx \gamma_d \frac{1 + L\dfrac{m_s}{R_d T}}{1 + L^2 \dfrac{m_s}{C_P R_v T^2}}$$

Appendix 6

Energy Equations – Derivation

We start with the equations of motion along the x and y directions.

$$\frac{\partial u}{\partial t} + u\frac{\partial u}{\partial x} + v\frac{\partial u}{\partial y} + \omega\frac{\partial u}{\partial p} = -\frac{\partial \phi}{\partial x} + fv + F_x \qquad \ldots (A6.1)$$

$$\frac{\partial v}{\partial t} + u\frac{\partial v}{\partial x} + v\frac{\partial v}{\partial y} + \omega\frac{\partial v}{\partial p} = -\frac{\partial \phi}{\partial y} - fu + F_y \qquad \ldots (A6.2)$$

Multiplying equations (A6.1) by u and equation (A6.2) by v and adding we obtain

$$\frac{\partial}{\partial t}\left(\frac{u^2}{2}\right) + u\frac{\partial}{\partial x}\left(\frac{u^2}{2}\right) + v\frac{\partial}{\partial y}\left(\frac{u^2}{2}\right) + \omega\frac{\partial}{\partial p}\left(\frac{u^2}{2}\right) =$$
$$-u\frac{\partial \phi}{\partial p} + fuv + uF_x$$

$$\frac{\partial}{\partial t}\left(\frac{v^2}{2}\right) + u\frac{\partial}{\partial x}\left(\frac{v^2}{2}\right) + v\frac{\partial}{\partial y}\left(\frac{v^2}{2}\right) + \omega\frac{\partial}{\partial p}\left(\frac{v^2}{2}\right) =$$
$$-v\frac{\partial \phi}{\partial p} - fuv + vF_y$$

Which yields

$$\frac{\partial k_H}{\partial t} + \vec{V}_3 \cdot \nabla k_H = -\vec{V}_H \cdot \nabla\phi + \vec{V}_H \cdot \vec{F}$$

where $k_H = \dfrac{u^2 + v^2}{2}$ is the horizontal kinetic energy, $\vec{V}_3$ is the three dimensional wind and $\vec{V}_H$ is the horizontal wind.

Integrating over a mass M

$$\frac{\partial K_H}{\partial t} = -\int \vec{V}_3 \cdot \nabla k_H dm - \int \vec{V}_H \cdot \nabla\phi \, dm + \int \vec{V}_H \cdot \vec{F} \, dm$$

It is interesting to note that the Coriolis force does not contribute to kinetic energy production as it is perpendicular to wind and does no work. It converts divergent component of wind into rotational component.

The main production (conversion from total potential energy) of horizontal kinetic energy is by the term

$$-\vec{V}_H \cdot \nabla\phi$$
$$= -\left(\vec{V}_g + \vec{V}_{ag}\right) \cdot \nabla\phi = -\vec{V}_{ag} \cdot \nabla\phi$$

where $\vec{V}_g$ is the geostrophic wind and $\vec{V}_{ag}$ is the ageostrophic, cross-isobaric wind.

Now, $$\vec{V}_g \cdot \nabla\phi = \frac{1}{f_0}\left(\hat{k} \times \nabla\phi\right) \cdot \nabla\phi = 0$$

As the geostrophic wind blows parallel to isobars it does not do work or contribute to production of kinetic energy.

It is the small ageostrophic or cross-isobaric wind that contributes to production of kinetic energy.

Alternatively, $\vec{V}_H = \vec{V}_\psi + \vec{V}_x = \hat{k} \times \nabla\psi + \nabla\chi$

It is the divergent component (velocity potential) that contributes to production of kinetic energy and not the rotational component.

In a low pressure area, if the isobars are circular, it is the radial component (inflow or outflow) that generates kinetic energy. In a tropical storm it is the toroidal circulation (inflow in the lower levels, upward motion inside and outflow in the upper levels and sinking outside) that sustains the cyclone. This is, of course, driven by the latent heat release.

In the Hadley circulation or monsoon meridonal circulation, it is the small meridonial component of wind that maintains the zonal wind (by Coriolis turning).

For getting an equation for the rate of change of total potential energy, we start with the thermodynamic energy equation.

Now, $$dQ = C_p dT - \alpha\, dp$$

i.e., $\quad \dot{Q} = C_p \dfrac{dT}{dt} - \alpha \dfrac{dp}{dt}$

$$= \dfrac{d}{dt}\left(C_p T\right) - \alpha\omega$$

Thus,

$$\dfrac{\partial}{\partial t}(P+I) = \int \dfrac{\partial}{\partial t} C_p T \, dm = +\int \dot{Q}\,dm - \int \vec{V_3}\cdot\nabla\left(C_p T\right)dm + \int \alpha\omega\,dm$$

List of Symbols used

p - pressure

ρ - density

α - specific volume
  - also angle, (depending on the context)

T – temperature

u – component of wind along the x-direction

v – component of wind along the y-direction

w – component of wind along the z-direction

x – distance (scalar) along the x-axis (positive towards the east)

y – distance (scalar) along the y-axis (positive towards the north)

z – distance (scalar) along the z-axis (positive upwards)

t – time

$\hat{i}$ - unit vector along the x-axis

$\hat{j}$ - unit vector along the y-axis

$\hat{k}$ - unit vector along the z-axis

k – wave number

λ - longitude

λ - also wavelength (depending on the context)

ϕ - latitude

ϕ - also geopotential (depending on the context)

c – phase speed

$\vec{V}$ - (vector) wind velocity

$\vec{F}$ - frictional force

R - gas constant

C_V - specific heat at constant volume

C_p - specific heat at constant pressure

ω - vertical p velocity

$\hat{t}$ - unit vector along the tangent

$\hat{n}$ - unit vector along the normal

ζ - vertical component of vorticity

ζ_a - absolute vorticity

ζ_p - potential vorticity

D – divergence

D_H - horizontal divergence

ψ - stream function

χ - velocity potential

$\vec{g}$ - acceleration due to gravity

g – gravity

f – Coriolis parameter

β - Rossby parameter

e – vapour pressure

L – Latent heat

  – also, wavelength, depending on the context

m – mixing ratio

  – mass (depending upon the context)

γ - lapse rate

γ_d - dry adiabatic lapse rate

$$\gamma = \frac{C_P}{C_V} \text{ (depending on the context)}$$

γ_s - saturation adiabatic lapse rate

$\vec{\Omega}$ - angular velocity of the earth

M – angular momentum

a – radius of the earth

σ - frequency

σ - static stability parameter (depending on the context)

Q – heating

θ - potential temperature

$\vec{r}$ - position vector

r – radius, radius of curvature

M – also mass (depending on the context)

$\vec{V}_g$ - geostrophic wind

V– scalar wind

V, τ – volume (depending on the context)

k – kinetic energy per unit mass (depending on the context)

K – kinetic energy

I – internal energy

P – potential energy

$$\kappa = \frac{R}{C_P}$$

References

Charney JG, 1947, The dynamics of long waves in a baroclinic westerly current: Jour. Met., vol. 4, pp 135-162.

Charney JG and Eliasssen A, 1964, On the growth of the hurricane depression: Jour. Atm. Sci., vol. 21, pp 68-75.

Charney JG and Stern ME, 1962, On the stability of the internal baroclinic jets in a rotating atmosphere: Jour. atmos. sci., vol. 19, pp 159-172.

Choudhury A.D, Krishnan R, Rama Rao M.V.S, Vellore R and Singh M, 2018, A phenomenological paradigm for midtropospheric cyclogenesis in the Indian summer monsoon: Jour. Atmos.Sci., vol.75, pp 2931-2954.

Dutton IA and Johnson DR, 1967, The theory of available potential energy and a variational approach to atmospheric energetics: Advan. Geoph., vol 12, pp 333-436.

Eady ET, 1949, Long waves and cyclone waves: Tellus, vol. 1, pp 33-52.

Frierson D.M.W, Lu J and Chen G, 2007, Width of Hadley cell in simple and comprehensive general circulation models: Geophysical Research Letters, vol 34(18).

Fultz D, 1951, Experimental analogies to atmospheric motions: Comp. Met, Am. Met. Soc., pp 1235-1248.

Gill A, 1980, Some simple solutions for heat induced tropical circulation: QJRMS, vol. 106, pp 447-462.

Goswami BN, Keshavamurty RN and Satyan V, 1980, Role of barotropic, baroclinic and combined barotropic baroclinic instability for the growth of monsoon depressions and mid tropophenic cyclones: Proc. Ind. Acad. Sci. (E & P segment)., vol. 89, pp 79-97.

Held, I. M and Hou, A.Y, 1980, Nonlinear axially symmetric circulations in anearly inviscid atmosphere: Jour, Atmos, Sci., vol.37, pp 515-533.

Hide R, 1953, Some experiments on thermal convection in a rotating liquid: Q.J.R.M.S., vol. 79, p 161.

Jeffreys H, 1926, On the dynamics of geostrophic winds, Q.J.R.M.S., vol 52, pp 85-104.

Kasture S.V., Satyan V and Keshavamurty RN, 1991, A model study of the 30-50 oscillation: Mausam, vol 42, pp 241-248.

Keshavamurty RN, 1982, Response of the atmosphere to sea surface temperature anomalies over the equatorial Pacific and teleconnections of the southern oscillation: Jour Atm. Sci., vol. 39, pp 1241-1259.

Keshavamurty RN, Asnani GC, Pillai PV and Das SK, 1978, Some studies of the growth of monsoon disturbances: Proc. Ind. Acad. Sci. (E & P Sci.)., vol. 87A, pp 61-75.

Kidson JW, Vincent DG and Newell RE, 1969, Observational studies of the general circulation of the tropics, long term mean values, Q.J.R.M.S., vol. 95, pp 258-287.

Kim Y-H and Kim MK, 2013, Examination of the global Lorenz cycle using MERRA and NCEP re-analysis 2: Climate dynamics, vol. 40, 5-6, pp 1499-1513.

Koteswaram P, 1958, The easterly jet stream over the tropics, Tellus A, vol. 10 (1), pp 43-57.

Kuo HL, 1949, Dynamic instability of two dimensional non-divergent flow in a barotropic atmosphere: Jour. Met., USA, vol. 6, pp 105-122.

Lorenz, EN, 1955, Available potential energy and the maintenance of the general circulation: Tellus, vol. 7, pp 271-281.

Manabe S and Smagorinsky J, 1967, Simulated climatology of a general circulation model with a hydrologic cycle II, Analysis of the tropical atmosphere: Mon. Wea. Rew., vol. 95, pp 155-169.

Marques CAF, Rocha A, Corte-Real J and Castanheira JM, 2009, Global atmosphere energetics from NCEP-Reanalysis 2 and ECMWF-ERA40 reanalysis: International Journal of Climatology, 29(2), pp 159-174.

Matsuno T, 1966, Quasi-Geostrophic motions in the equatorial area: J Met Soc., Japan, vol. 44, pp 25-42.

Miller, F.R and Keshavamurthy R.N, 1968, Structure of an Arabian sea summer monsoon system: Meteorological monograph, East-West Center Press, Hawaii.

Palmen EH, 1964, General circulation of the tropies: Proc. Sym. Trop, Met, Rotorua, New Zealand, Nov. 1963, pp 3-30.

Pfeffer RZ, Mardon D, Sterbenz P and Fowlis W, 1965, A new concept of available potential energy: Proc. Int. Sym. Dynamics of large scale atmos. Processes., Moscow, pp 179-189.

Phillips NA, 1954, Energy transformations and meridional circulations associated with simple baroclinic waves in a two-level quasigeostrophic model: Tellus, vol. 6, pp 273-286.

Phillips NA, 1956, The general circulation of the atmosphere – A numerical experiment: Q.J.R.M.S., vol. 82, pp 123-164.

Pisharoty PR, 1954, The kinetic energy of the atmosphere, Sci. Rep. No. 6, Uni. of California, Los Angeles, Dept. of Met., Contract AF 19(122)-48, pp 140.

Rao KN, 1958, Thermodynamical instability of the upper atmosphere over India during the southwest monsoon: Proc. Sym. Monsoons of world, New Delhi, pp 43-52.

Satyan V, Keshavamurty RN, Goswami BN Dash SK and Sinha HSS, 1980, Monsoon cyclogenesis and large scale flow patterns over south Asia: Proc. Ind. Acad. Sci. (E & P Sciences)., vol. 89, pp 277-292.

Smagorinsky J, Manabe S and Holloway J, 1965, Numerical results from nine level general circulation model of the atmosphere : Mon. Wea. Rev., vol. 93, pp 769.

Starr VP and White RM, 1951, A hemispherical study of the atmospheric angular momentum balance: Q.J.R.M.S., vol. 77, pp 215-225.

Starr VP, 1968, Physics of negative viscosity phenomena: (Earth and planetary science series).

Swapna P, Krishnan R, Sandeep N, Prajeesh AG, Ayantika D.C., Manmeet S and Vellore R, 2018, Long term climate stimulations using the IITM earth system model (IITM – ESMv2) with focus on the south Asian monsoon: Journal of advances in modeling earth systems, AGU 100.

Van Mieghem J, 1956, The energy available in the atmosphere for conversion into kinetic energy: Beitr. Phys. Fr. Atmos., vol. 29, pp 129-142.

Wiin-Nielson A, 1968, On the intensity of the general circulation of the atmosphere: Rev. of Geoph., vol. 6, pp 559-579.

Bibliography

Halltiner GJ and Martin FL, 1957, Dynamical and physical meteorology, McGraw Hill book company.

Holton JR, 1992, An introduction to dynamic meteorology, Acad. Press Inc., International Geoph. Series.

Haltiner GJ, 1971, Numerical weather prediction, John Wiley and sons.

Keshavamurty RN and Sankar Rao M, 1992, The physics of Monsoons, Allied Publishers Ltd.

Pedlosky J, 1979, Geophysical Fluid Dynamics, Springer Verlag.

Peixoto JP and Oort AH, 1992, Physics of climate, American Institute of Physics.

For Product Safety Concerns and Information please contact our EU
representative GPSR@taylorandfrancis.com Taylor & Francis Verlag GmbH,
Kaufingerstraße 24, 80331 München, Germany

Printed and bound by CPI Group (UK) Ltd, Croydon, CR0 4YY
01/05/2025
01858560-0001